Writing Virtual Environments for Software Visualization

Clinton Jeffery • Jafar Al-Gharaibeh

Writing Virtual Environments for Software Visualization

Clinton Jeffery
Department of Computer Science
University of Idaho
Moscow
ID
USA

Jafar Al-Gharaibeh
R&D Department
Architecture Technology Corporation
Eden Prairie
MN
USA

ISBN 978-1-4614-1754-5 ISBN 978-1-4614-1755-2 (eBook)
DOI 10.1007/978-1-4614-1755-2

Library of Congress Control Number: 2014957537

Springer New York Heidelberg Dordrecht London

Printed on acid-free paper

Springer is part of Springer Science+Business Media (www.springer.com)

Preface

This book combines two emergent research areas of the computer graphics world. On the one hand, there is software visualization. After 30 years, it remains an under-achiever with enormous potential to impact how we develop software. This book looks at how to visualize dynamic program behavior, while it is happening. The techniques are readily applicable to other areas of software visualization. Our ultimate goal is to make visible many practical aspects of program behavior that are currently invisible or difficult to see at best, such as how to find bugs and performance bottlenecks.

A second major focus of this book is on virtual environments. Lots of programmers want to create them, but for most that challenge is just *too* hard; virtual environments integrate advanced 3D graphics, animation and networking in ways that most ordinary developers can't manage in a practical time frame. We are writing this book to do what we can to help conquer that obstacle.

It is obvious that these two genres of computer graphics should be combined. The networking and persistence afforded by virtual environments are exactly what software visualization needs in order to become more collaborative, shared, ubiquitous, and educational. All we have to do is figure out how to build it.

Unicon is an innovative, very high-level programming language hosted at unicon.org and the popular open source site SourceForge. It is descended from Icon, a language developed at the University of Arizona as a successor to the SNOBOL language family from AT&T. Unicon's 3D and networking facilities turn out to be great for virtual environments, and its execution monitoring facilities make it second to none in the area of software visualization. Programs in other languages can often be instrumented to extract similar program behavior information, but in Unicon 120 or so kinds of program execution behavior are continuously available, any time they are of interest.

This book is organized into two parts. Part I is an overview of execution monitoring and program visualization. It presents Unicon's monitor architecture and the framework for monitoring Unicon programs using a series of example visualization tools that observe many kinds of execution. Part II presents virtual environments, including 3D modeling and network communication necessary for multi-user collaboration. It introduces methods of propagating visualization information and collaborating within a multi-user virtual environment. The ultimate goal is to produce game-like real-time graphical ways of understanding and talking about bugs, bottlenecks, and other aspects of program behavior. Following Part II is a collection of appendices including some detailed program examples and a description of the implementation of the monitoring framework.

Acknowledgments

Intellectual contributions to our work were made in the University of Idaho and New Mexico State University graduate courses on program monitoring and visualization, and on collaborative virtual environments. The authors wish to thank the students from those courses.

This book descends from the 1999 tome *Program Monitoring and Visualization*, whose acknowledgements section recognizes Ralph Griswold, Rick Snodgrass, Mary Bailey, Norm Hutchinson, Wenyi Zhou, Kevin Templer, Gregg Townsend, Ken Walker, Michael Brazell, Darren Merrill, Mary Cameron, Jon Lipp, Nick Kline, Song Liang, Kevin Devries, Steve Wampler, Thanawat Lertpradist, Laura Connor, Anthony Jones, Khan Mai, and Niem Tang.

Brett Kurzman of Springer is responsible for initiating this book. Mary James and Rebecca Hytowitz at Springer provided the kind of patience and support that all authors want.

This work was supported in part by the National Science Foundation under grant DUE-0402572, and by the National Library of Medicine Division of Specialized Information Services.

Moscow, Idaho and Eden Prairie,
Minnesota, September 2014

Clinton L. Jeffery and
Jafar M. Al-Gharaibeh

Contents

List of Figures

List of Tables

Chapter 1
Introduction

As the demand for computer software grows to include more diverse and larger applications, software developers desperately need better tools to aid in the understanding of dynamic aspects of program behavior during various phases of the software life cycle, including debugging, performance tuning, and maintenance.

Spectacular improvements in graphics, concurrency and networks during the past decade have put us in a position where we ought to be able to see what is going on while our programs are running, in rich, animated 3D, and to easily share and consult with each other while studying such visualizations of program behavior. It ought to be no harder than, say, playing *World of Warcraft*.

This book applies videogame technology to software development by making the vision of multi-user online collaborative 3D software visualization a reality. The topic will be explored by demonstrating an example implementation, getting down into its code.

1.1 Software Visualization and Program Behavior

The reason to want software visualization is to gain a better understanding of a program's behavior. Some program-understanding systems convey very specific information about a small portion of a program, such as the workings of a single algorithm. Others are concerned with explaining the role that a program or a collection of programs plays within a larger computational system. This book addresses a common problem in between these two extremes: understanding the workings of a single (possibly large) program.

People who need to understand a program usually have two alternatives: studying the source code, or running the program to see what it does. Ideally, a program would be understandable using one or the other of these methods; in practice, reading source code is impractically cumbersome for many programs, and construction of test cases to explain program behavior is often a tedious and speculative undertaking. These difficulties motivate the development of special programs that are used to help explain the behavior of other programs.

C. Jeffery, J. Al-Gharaibeh, *Writing Virtual Environments for Software Visualization*, DOI 10.1007/978-1-4614-1755-2_1

Program-understanding systems are used in a variety of applications. The most common motive for program-understanding is *debugging*. Programs that produce incorrect output or fail to complete their execution due to bugs are prime candidates for tools that assist program developers and maintainers in program-understanding tasks. A *debugger* is a program designed specifically to help with the debugging process. General-purpose program-understanding tools are also used to assist in *debugging*.

A second major application of program-understanding systems is *performance tuning* or *performance debugging*. A correct, working program may be of limited usefulness if its performance is poor. Frequently, a program's authors or maintainers can improve execution speed by using different programming techniques or modifying the program's algorithms and data structures. By providing an accounting of which resources the program is using and which sections of code are primarily responsible, performance tuning systems can direct programmers' efforts to where they are most needed.

A third application of program-understanding is *software instruction* and *orientation*. The internal workings of a program may be of special interest to students learning important algorithms, data structures, or programming techniques; this situation frequently arises when learning a new language. People assigned to maintain or improve a program written by someone else similarly need to *orient* themselves as to its general operation. In both of these cases, the people involved may be entirely unfamiliar with the program source code, and can benefit from information provided by program understanding tools before consulting source code, or without referring to it at all.

In addition to these established uses for program-understanding systems, program-understanding tools can provide language implementors with valuable assistance in the task of *language implementation tuning*. Program-understanding tools that provide information about the execution of programs also directly or indirectly provide information about the language's implementation. This information can be used to improve performance or address problems in the implementation.

1.2 Types of Software Visualization Tools

Programs that provide information about other programs can be separated into two main categories based on the kind of information they provide. *Static analysis tools* examine the program text and, in conjunction with knowledge of the language, provide information about a program that is true for all executions of that program independent of its input [1]. Compiler code optimizers, pretty printers, and syntax-directed editors frequently employ static analysis techniques. This book employs static analysis mainly to provide a spatial context, the terrain within which a user interprets dynamic behavior, perceived as animated objects.

Dynamic analysis tools provide information about a specific program execution on a specific set of input data [1]. Since dynamic analysis involves extracting information from a running program rather than its source code, these tools pose implementation problems that are very different from those found in static analysis tools. Another name for a dynamic analysis tool is a *program execution monitor.* A program execution monitor is a program that monitors the execution of another program [2]. Program execution monitors complement static analysis tools and provide execution information that static tools cannot, such as details about the program's control flow, intermediate results that are computed, or depictions of internal data structures as the program runs. On the other hand, static aspects of a program such as variable names often provide context crucial to the understanding of execution behavior. Good dynamic analysis tools incorporate static information in support of dynamic information. Execution monitors include the source-level debuggers and profilers commonly bundled with compilers and available on many operating systems.

An execution monitor may either present information to the user as the program executes (immediate or *runtime* analysis), or it may present information at some later time such as after execution completes (*postmortem* analysis). Runtime analyzers provide immediate feedback and allow user direction of the kind and level of detail of the information monitored. In contrast, postmortem analyzers may perform extensive computations to condense the execution information and present it in a useful way. The two methods are not mutually exclusive.

Runtime analysis tools can further be categorized as *passive* or *interactive.* In a passive system, the tool presents information to the user, but the user has little control over the activity. In an interactive system, the user may have external control over what information is displayed, or even may be able to modify the computation being observed or the data being processed.

1.3 Virtual Environments

For the purposes of this book, a virtual environment is an online, multi-user 3D space within which users share activities and information. Although popularized by videogames, virtual environments were described long ago in books such as Gibson's *Neuromancer.* We are simply taking that concept and applying it for a specific serious application in a manner that makes sense given contempory hardware and software capabilities. That said, we hope to advance the state of the art and go where no one has gone before, and make it a lot of fun, too.

1.4 Scope of This Book

This book uses Unicon's built-in execution monitoring facilities to obtain the information to be visualized. Monitors employ Unicon's high-level 2D and 3D graphics facilities for visualization tasks. Things get interesting when these tools are embedded within a software system that enables them to be easily shared across the internet, within a game-like 3D virtual environment.

This book discusses execution monitors within a well-defined context: the Unicon programming language. Unicon is a high-level object-oriented language that descends from Icon and SNOBOL4.

The way in which a program uses structured data types, algorithms, and function or class libraries has a fundamental effect on program execution behavior. In Unicon these behaviors can be observed via the language's built-in, primitive operations and runtime system behavior.

The primary tasks of an execution monitor are to collect information about a program's execution and present that information to the user in an understandable way. In addition to the inherent complexity of these tasks, the main problems posed by execution monitoring are:

- **Volume**: The large amount of data to be processed by the monitor code entails performance problems both in the gathering of information and in the presentation of that information. Efficient gathering of information involves selecting the relevant information from the huge pool of available program behavior data. Efficient presentation of information includes making effective use of the visual medium to communicate with the user, as well as understanding the user's powers of perception. Even if it is gathered and presented efficiently, the large amount of information inherent in monitoring tends to obscure items of interest.
- **Dimensionality**: Program execution, although it can be viewed as a sequence of steps executed by the computer, is more completely characterized as a trajectory of a point moving through *n*-dimensional space [2]. For example, at any given instant during execution, an understanding of program behavior can depend on information about current stack depth, source location, heap activity, input/output, and so forth. Each programming language has its own additional dimensions of observable behavior corresponding to its semantics. In addition to behavior related directly to explicit source program operations, very high-level languages also have significant dimensions of implicit runtime system behavior, such as automatic memory management and type conversions.
- **Intrusion**: All monitoring systems alter the execution environment of the program under study; when the act of monitoring a program changes the behavior under observation, it is called *intrusion* [3, 4]. Henry defines *control-intrusive* and *data-intrusive* methods of adding instrumentation to a program in order to monitor its execution [4]. Control-intrusive instrumentation takes the form of code (such as a procedure call to a monitor routine) embedded within the program. Data intrusion arises in object-oriented systems in which instrumentation is added by subclassing a class to be instrumented and overriding its access methods with additional code. The subclass calls monitor code in addition to calling the superclass method(s) to perform the normal computation. The term intrusion has also been used to refer to the execution slowdown imposed by monitoring [3]; in realtime and concurrent systems, this can render monitoring useless. This form of intrusion is not considered in this work. The effect of monitoring on

execution speed is considered only so far as to establish framework viability on real Unicon programs.

- **Access**: Execution monitors often require extensive access to the variables and structures in the program being monitored. If the monitor and program being monitored are distinct programs, operating system constraints may restrict this access, or create performance problems in this area, or both. From the point of view of the execution monitor author, the access problem may also be reflected by low-level or cumbersome notations used to read or write target program data. A good example of access is the traversal of pointers in data structures: if it requires operating system intervention or a notation other than that used in the target program source code, the monitor has poor access to the target program, and the task of writing monitors is made difficult. Solutions to the access problem, such as adding monitor code directly to the program being monitored, often aggravate the intrusion problem.

These problems are universal in execution monitoring and appear repeatedly in the literature. While no general solution for these problems exists, improved monitoring techniques may lessen their severity or solve them for practical purposes on real programs. Traditionally the implementation of execution monitors has been very difficult because the programmers implementing a new monitor necessarily spent a considerable effort addressing these problems. The difficulty of implementing monitors in turn limits or effectively prevents efforts to improve monitor technology by experimental means.

1.5 Contributions

The goal of this book is to reduce the difficulty of constructing execution monitors and virtual environments by developing a practical framework in which monitor construction is relatively easy, virtual environment construction is relatively easy, and embedding a monitor within a virtual environment is no big deal. The problems identified in the previous section motivate the chosen solutions.

This book provides source language support for the central act of gathering execution information. It addresses the problems of volume, dimensionality, intrusion, and access in the following ways:

- Built-in language features for the central act of gathering execution information provide the performance that is necessary for effective monitors written in the source language, despite the generally slower speed of very high-level languages. Dynamic control over the information flow from the program to the monitor is essential for performance.
- Multiple monitors simultaneously observe the same program execution and present views along different dimensions of program behavior. Support is provided for monitor communication and the exploratory construction of monitors that specialize in monitor coordination.

- Language support for gathering execution information from the runtime system eliminates code intrusion. Provision of separate memory allocation areas for the monitor and target program avoids data intrusion.
- Source language support allows the execution of the monitor and target program in a shared interpeter and provides full source-level access of the monitor to the target program. The framework uses a synchronous coroutine execution model and a shared address space, offering significant advantages without restricting the kinds of monitors that the system supports.

In addition to these features that address core execution monitoring tasks, the framework provides full separation of the program and the various monitors that observe it. Taking the form of dynamic loading and a virtual monitor interface, this separation provides the ease of use that is necessary in order to provide exploratory programming capabilities. The separation allows multiple monitors to observe a program at the same time, and allows new monitors to augment or enhance the capabilities provided by existing monitors.

The intent of the framework is to provide shareable exploratory visualization programming capabilities not just for experts, but also for applications programmers who are trying to understand their programs. Given this framework and appropriate library support procedures, writing a new visualization tool is no more difficult than writing other applications that involve communication between programs, and often is simpler than writing such applications.

References

1. R. Dunn, Software Defect Removal, New York: McGraw-Hill Book Company, 1984.
2. B. Plattner and J. Nievergelt, "Monitoring Program Execution: A Survey," *IEEE Computer*, pp. 76–93, #nov# 1981.
3. Z. Aral and I. Gertner, "Non-intrusive and Interactive Profiling in Parasight," in *Proceedings of the ACM/SIGPLAN PPEALS 1988*, 1988.
4. R. R. Henry, K. Whaley and B. Forstall, "The University of Washington Illustrating Compiler," in *Proceedings of the ACM SIGPLAN '90 Conference on Programming Language Design and Implementation*, White Plains, NY, 1990.

Part I
Software Visualization

Chapter 2
Visualization Principles and Techniques

The previous chapter described many monitoring systems' execution models and methods of extracting behavior information from programs that they monitor. These "hard" tasks are the primary subject of this book, but observing behavior is pointless unless information is presented to the user in a way that meets his or her needs. Visualization is the art of presenting large amounts of information in accessible graphical form. Appropriate graphic representations allow the rapid delivery of vast amounts of information that would overwhelm the user if presented in textual form. This chapter introduces the principles that underline effective visualizations, suggests an array of visualization techniques, and presents an incremental methodology by which such tools may be developed.

It is worth noting that there is more than one kind of visualization. *Scientific visualization* is the graphic rendering of gigantic *n*-dimensional data sets by various means of projection and abstraction. In contrast, *software visualization* is the depiction of software artifacts such as directories, user data, or system log files.

Program visualization as described in this book is a subfield of software visualization focused on the dynamic behavior of programs themselves, rather than the data they manipulate. The preceding chapter did not present broad coverage of related work in the general area of software visualization, since it has been described admirably elsewhere [1].

Compared with scientific visualization, program visualization is more abstract, since program behavior above the hardware level does not map easily onto real-world geometries. Program visualization evolved from the hand-written diagrams and notations used by programmers and computer scientists to describe their structures prior to the advent of automated visual tools.

Before there was visualization, there was graphic design. Visualization emerged as a subdiscipline of graphic design when computer screens began to replace printed paper. Visualization includes graphic design, but has additional constraints imposed by hardware and software capabilities and user requirements.

C. Jeffery, J. Al-Gharaibeh, *Writing Virtual Environments for Software Visualization*, DOI 10.1007/978-1-4614-1755-2_2

2.1 Graphic Design Principles

It is not worth implementing elaborate computer graphics if the graphic design does not convey information clearly. Some principles of graphic design are self-evident, such as abstracting away irrelevant detail; other principles are learned through experience. Some of the best references on graphic design are by Tufte [2–4] and Bertin [5]. Tufte's observations concerning graphic excellence are summed up by the following:

Graphical excellence is that which gives to the viewer the greatest number of ideas in the shortest time with the least ink in the smallest space.

To achieve such excellence in designs, Tufte advocates five principles:

- if you do nothing else, at least show the information
- show as much as you can with as little ink as possible
- remove ink that isn't showing useful information
- remove redundant information
- revise and edit

The reader may consult Tufte's work for numerous examples of these principles in practice.

2.1.1 Show the Information

If a graphic fails to deliver the required information, it is worse than no graphic. While this mistake might not sound likely, it is not so uncommon for a visualization to present only part of the required information. Depending on the context this can be misleading, or it can be disastrous. Omitting critical information may be as serious as distorting it, a violation of this principle that takes place in myriad forms [6].

Perhaps the most common cause of lost information is when some other information is mapped onto the same location and obscures an item of interest. The viewer may not even know that something is covered up. The situation is not much better if the information is presented but the user cannot decipher it. Visualizations should be self-explanatory. Axes and units, both for geometry and the temporal interpretation of animations, should be clearly explained. On a print graphic, labels and legends might pose an obstacle themselves, but on a computer display they can be toggled on when needed and disappear just as easily.

2.1.2 Maximize Information Density

A high information density allows more information in the available space, or a given amount of information in a smaller space. Tufte gives published examples that range over between two and three orders of magnitude from the least dense to the most dense! Many computer visualization tools are produced by computer scientists who are experts in fields such as architecture or parallel algorithms and do not bother to study graphic design before they publish their visualization work; not surprisingly, some of these tools achieve monumentally low information density.

Information density is increased by eliminating empty space by either adding more information or shrinking the display. Only the actual information depicted counts toward high density, not labels, legends, axes, arrows, or notes present. If you shrink your display not to the point of unreadability, but to the minimal size it requires, you will make room to run multiple visualization tools side by side. The architecture described in this book supports a large number of simultaneous visualization tools.

High information density comes at a price. Humans have a minimal resolution of perception. Humans can see individual pixels on modern displays, but to perceive discrete objects animated on a screen, the minimal comfortable size may vary, depending on the resolution and size of the display. Don't rely too heavily on an individual pixel to make an important point.

Information density should be increased in a way that keeps the graphic organized. A dense, complex graphic may be so cluttered that it is difficult to sort out. It turns out that *organized complexity* allows rapid assimilation and is aesthetic in appearance, while *disorganized complexity* is difficult for humans to handle [7].

2.1.3 Remove Useless Information

Many or most visualization tasks have more information available than room in which to display it. Even if there is enough room, nonessential information distracts and detracts from important information. The user has only so much attention to go around. For this reason, filtering out useless information is a basic task. Of course, not all information can be designated as useless or useful; information occupies a range of utility, especially in visualization situations in which the users are not sure what they are looking for. A corollary of the principle of removing useless information is that the emphasis (and space) allocated to information should be proportional to its usefulness.

Consider drawing common tree structures where many nodes have parent-child relationships, and each node contains its own information. Figure 2.1 shows such a tree using a classically attractive layout algorithm due to Moen [8]. This layout is

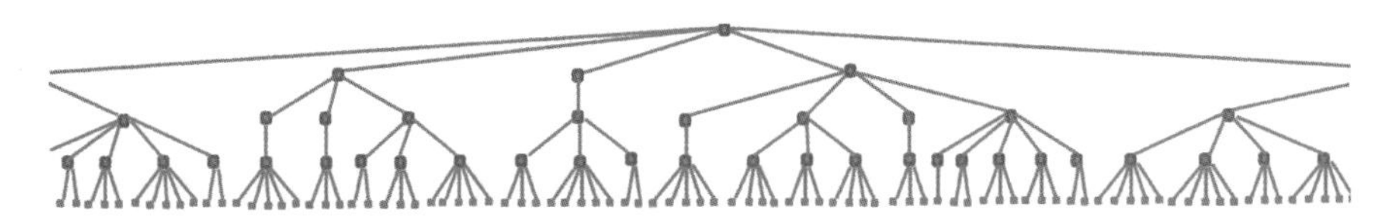

Fig. 2.1 A classical tree layout

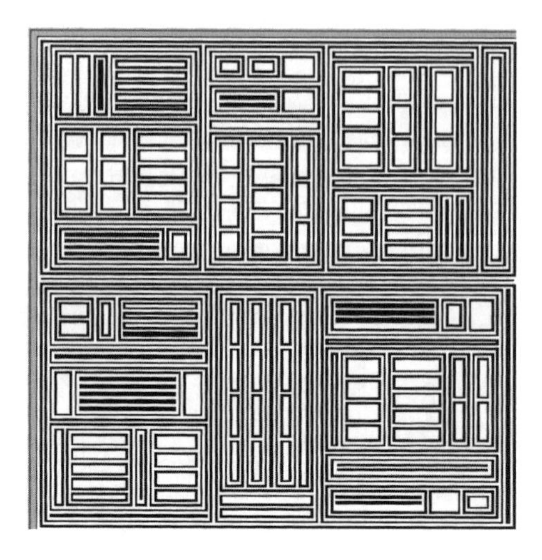

Fig. 2.2 A tree-map emphasizes nodes instead of edges

fine if the user is mainly interested in the size and shape of the tree structure itself; almost all screen space is devoted to structure.

In many applications, however, the user may be more interested in the information presented at the nodes; the tree structure may be a side issue. In this case, a layout such as the tree-map [9] shown in Fig. 2.2 may be more relevant, even though its portrayal of the tree structure is less clear. The layout you select for drawing trees should depend on what is important to the user at that particular moment.

2.1.4 Remove Redundant Information

In user-interface design, it is good for users to have many different ways to accomplish whatever they are trying to do. It is similarly good to have several different ways of viewing data—one way at a time. Providing several simultaneous views of the same data is wasting scarce pixels and resolution that can be better applied to providing a single, richer view. This is even more true for visualization than for ordinary graphics, since visualization typically involves abstracting away a richer set of data than can be portrayed in detail. Useless information is bad, and so is duplicated information.

2.1.5 Iterate

It is naive to expect that your first approach to visualizing something will be optimal. The graphic design will evolve. These refining iterations are costly and time consuming for printed graphics, and they can be equally difficult for visualization programming. Using a language and graphics toolkit that accommodates this principle is common sense.

2.2 Visualization Principles

Most visualization efforts can start by adapting a well-known technique from printed graphics. Where graphic designers say *ink* one may generally substitute *pixel writes* to the computer's display. A typical bitmapped computer graphics display presents a million or more bytes of information to the user at a time. The raw number of pixels and colors available defines the capability of the display to render static images analogous to print graphics.

Static computer images can be evaluated in terms of graphic design principles such as information density. Most visualizations, however, are dynamic; the graphics hardware and software define limits on the rate at which the display may change. In this context, information density is a three-dimensional measure that includes the time interval during which the changing display is viewed.

Visualization of dynamic execution behavior is different from visualizing a static data set in several ways. These differences motivate the techniques presented in the rest of this book. They may be summarized in the following basic concepts:

- animation
- metaphors
- interconnection
- interaction
- dynamic scale
- static backdrop

2.2.1 Animation

The ability to depict temporal relationships by animating dynamic behavior is a crucial tool. There are tradeoffs between visual sophistication and the associated computational cost and programming time required. Widely applicable techniques are ones that can be animated on low-cost hardware.

Thanks to the computer games market, our definition of low-cost hardware now includes support for animating 3D scenes of moderate complexity. While ubiquitous, many specific workstations and operating systems do not support 3D graphics, so they are not universally available. The hardware is not really the main problem.

The main obstacle to widespread development of animated 3D visualizations is inadequate software. OpenGL, the closest thing to a universal standard, is a low-level specification more oriented towards hardware capabilities than ease of programming. Higher-level languages and toolkits for 3D programming are nonportable, expensive, or both.

In order to achieve universally available, easily programmed animations, this book focuses on simple 2D graphic designs. The output of monitoring could be piped into an existing visualization package such as IDL or Khoros for more sophisticated graphics rendering including 3D views, but this would reduce the degree of interactivity and control provided by the tools.

2.2.2 Least Astonishment

Visualizations should obey the principle of least astonishment. This is important in print graphics, but it is even more important in animated visualizations where the display is changing and the users cannot continuously study each image at their leisure. When possible, visualizations should present information in a manner with which the users are already familiar.

Tufte observes that in the absence of a reason to do otherwise, most graphics should utilize the *golden rectangle*, with a primary horizontal axis 1.6 times wider than the vertical axis. The arguments in favor of the golden rectangle range from human psychology to alleged evolutionary skill at scanning the horizon for predators and prey. Computer displays favor the horizontal axis. Text labels are read horizontally and are understood more rapidly when written in a single line than when split onto multiple lines, again favoring a primary horizontal axis. Because most graphics depict information similarly, presentation of data will produce less astonishment when the horizontal axis represents the *cause*, and the vertical axis represents the *effect* described by the graph.

2.2.3 Visual Metaphors

The mapping from program information to window geometry often is artificial or unintuitive, especially when no natural geometry is inherent in the information to be presented. A familiar or readily inferred visual metaphor for the behavior being presented can lower the cognitive load imposed on the user and increase the rate of comprehension.

Although some metaphors are drawn naturally from a specific application domain or a notation in common use among programmers, others are drawn from nature or from nontechnical symbols found in daily life.

2.2.4 Interconnection

Understanding a complex piece of software entails an understanding of a variety of distinct behaviors and the relationships between them. For example, control flow, data structures, memory allocation behavior, and input/output all have distinct but interrelated patterns in program execution. Visualizations that consume most or all of the screen do not allow for simultaneous display of other forms of execution behavior.

Under any circumstances, we cannot hope to portray all of program execution in a single 2D or 3D graphic. And we cannot anticipate, in general, which subset of the available information a given user will need. Our emphasis is on multiple visualization tools selected by the user, executing simultaneously, showing different aspects of program behavior.

2.2.5 Interaction

Visualizations are more effective when the user can navigate and steer them in appropriate directions. Operations such as panning and zooming are vital in order to present some information in more detail than others, because the user remains vital in prioritizing which information to emphasize. The importance of navigation increases in 3D visualizations. Depictions of 3D objects on a computer screen may be ambiguous unless those objects are seen in motion.

A graphic design used in visualization should allow for natural interactive controls, an issue not addressed in static design. For example, a user should be able to select objects and request details about them, or specify that they should be watched, and execution should pause when they are modified.

2.2.6 Dynamic Scale

The scale imposed in the depiction of dense information on a computer sceen is extreme, but in addition, the scales are highly dynamic. If the scale does not change dynamically, most visualizations waste space and lose detail over most of the execution being observed. On the other hand, changing scale too frequently is both computationally expensive and disorienting.

There are scaling alternatives to redrawing the entire window to use larger or smaller units. There are several different ways to utilize a scale that varies in a single image in a consistent way. If one of these techniques is used, it must be evident to the user or it can work harm instead of help. Logarithmic scales are one option, but they are not always appropriate and typically need to be tuned to the size of the dataset involved.

A more generally useful, but treacherous dynamic scaling technique, is the fisheye view [10]. Fisheye views introduce one or more focus points of attention; a distortion is applied to all output to the window, scaling it by some function of its distance from the focus or foci. Figure 2.3 shows a simple fisheye view with a single focus point applied to a text file. A scalable font and simple arithmetic were all that was needed in order to render this view.

Fisheye views can be applied to arbitrary graphics, and distortion functions may account for multiple focus points with varying degrees of importance [11]. Figure 2.4 shows a 2D map of a simple graph in which each node is distorted by its own factor. The nodes in this case represent solar systems, and they are annotated by orbiting planets and text labels that are all scaled proportional to the size of the star itself, which is drawn as a yellow circle. Planet details and text labels are omitted for systems drawn beneath a minimum threshold size.

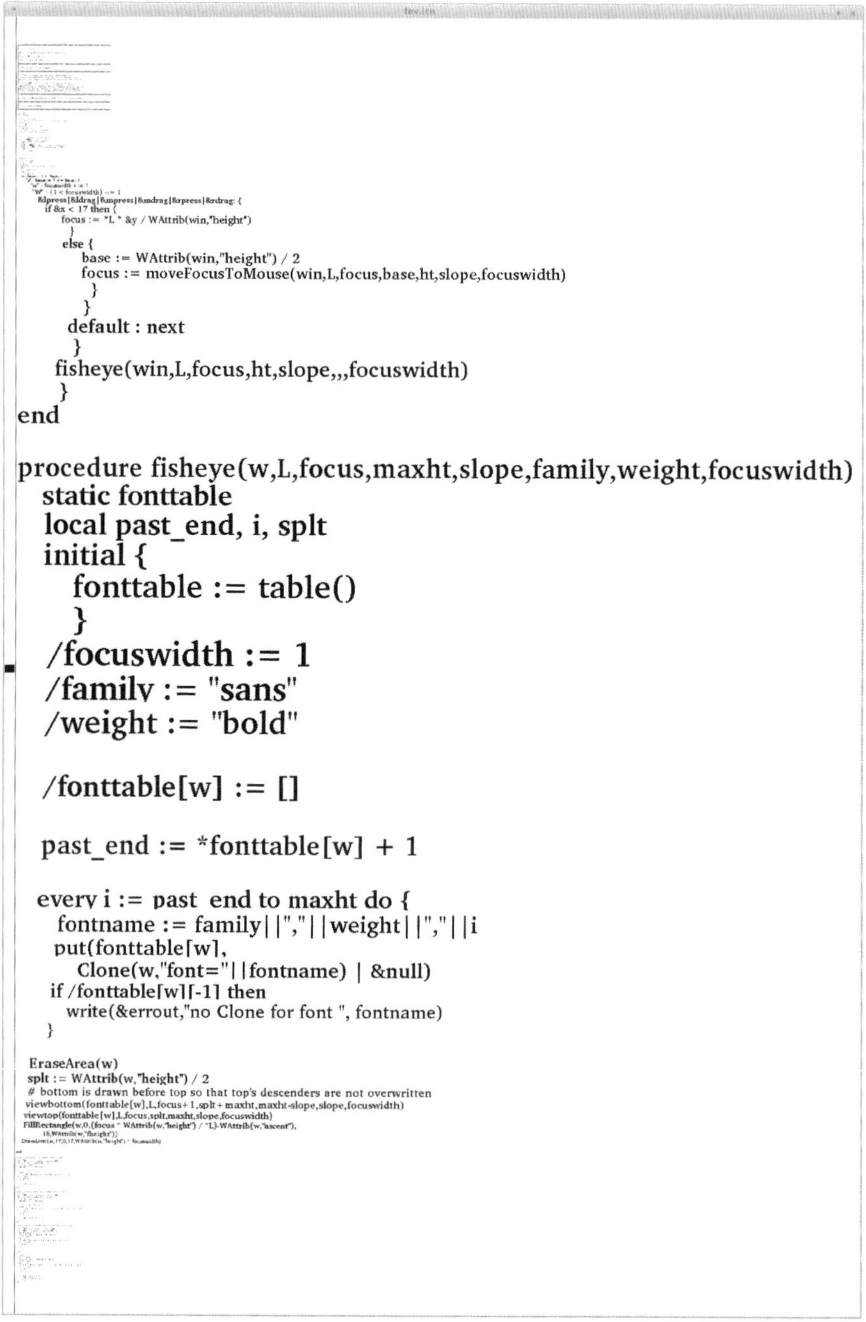

```
    &lpress|&ldrag|&mpress|&mdrag|&rpress|&rdrag: {
      if &x < 17 then {
          focus := *L * &y / WAttrib(win,"height")
          }
        else {
          base := WAttrib(win,"height") / 2
          focus := moveFocusToMouse(win,L,focus,base,ht,slope,focuswidth)
          }
        }
      default : next
      }
    fisheye(win,L,focus,ht,slope,,,focuswidth)
    }
end

procedure fisheye(w,L,focus,maxht,slope,family,weight,focuswidth)
   static fonttable
   local past_end, i, splt
   initial {
      fonttable := table()
      }
   /focuswidth := 1
   /family := "sans"
   /weight := "bold"

   /fonttable[w] := []

   past_end := *fonttable[w] + 1

   every i := past_end to maxht do {
      fontname := family||","||weight||","||i
      put(fonttable[w],
         Clone(w,"font="||fontname) | &null)
      if /fonttable[w][-1] then
         write(&errout,"no Clone for font ", fontname)
      }

   EraseArea(w)
   splt := WAttrib(w,"height") / 2
   # bottom is drawn before top so that top's descenders are not overwritten
   viewbottom(fonttable[w],L,focus+1,splt+maxht,maxht-slope,slope,focuswidth)
   viewtop(fonttable[w],L,focus,splt,maxht,slope,focuswidth)
   FillRectangle(w,0,(focus * WAttrib(w,"height") / *L)-WAttrib(w,"ascent"),
         16,WAttrib(w,"fheight"))
   DrawLine(w,17,0,17,WAttrib(w,"height") * focuswidth)
```

Fig. 2.3 A fisheye view with a single focus point

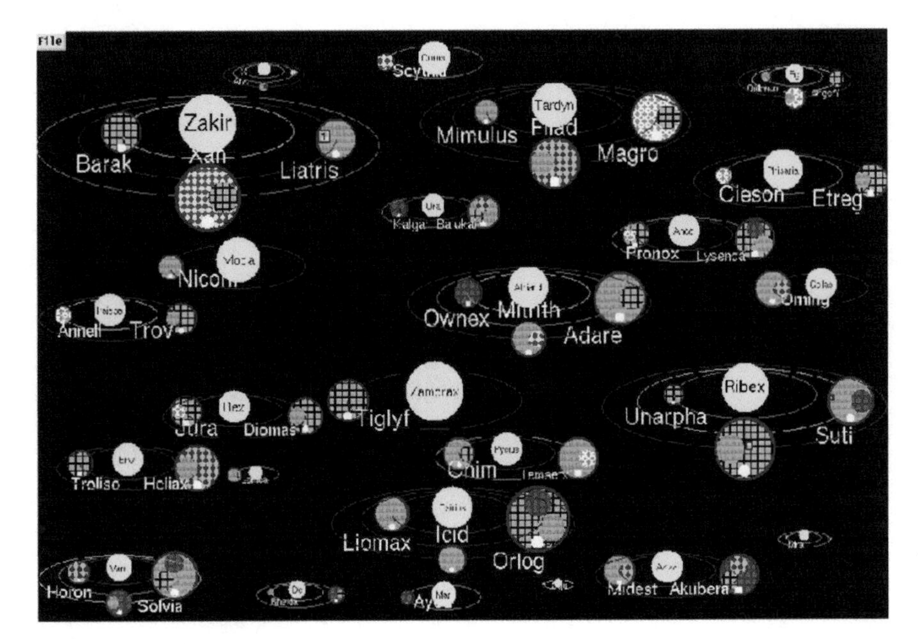

Fig. 2.4 A graphical fisheye view

2.2.7 *Static Backdrop*

Dynamic analysis tools are often best interpreted when superimposed upon a context consisting of information acquired by static analysis; the static information can provide a map that programmers are familiar with. Examples of static backdrops are a program's call graph, or even its source code.

2.3 Visualization Techniques

The visualization author faces the problem of rendering the selected graphics with an acceptable real-time performance, characterized by animation frame rate as well as interactive responsiveness and navigability. A visualization can be measured as graphics hardware is measured, in pixels or polygons per second, or the frame rate at which the screen is updated. While this may tell you whether a workstation with faster graphics would improve your visualization, it says nothing about whether the user understood the information. Humans cannot follow details very rapidly, so the optimal rate of change for the graphic display is more likely to be bounded by the human reader than by the hardware or the visualization's graphic rendering algorithm.

Fortunately, some of the simplest graphics are effective, easy to implement, and are familiar to users. Time series graphs, bar charts, pie charts, and scatterplots are all examples of graphic designs that are easily programmed but may need adapta-

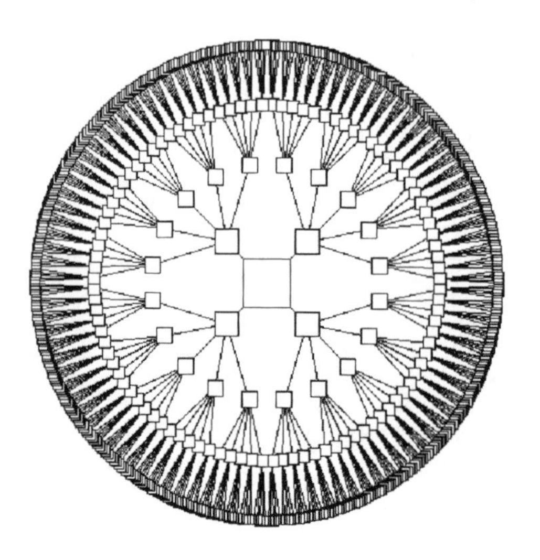

Fig. 2.5 A circular tree

tion for visualization purposes. Part III of this book includes many examples of such adaptation.

2.3.1 *Incremental Algorithms*

Redrawing the entire screen each time something changes is not an efficient approach. Incremental algorithms may be required in order to achieve smooth animation. An incremental algorithm is smart enough to render graphics only for the objects on screen that are affected by an operation, and not redraw the others.

It is easy to write incremental algorithms for some graphic designs, and not easy for others. The importance of speedy animation often dictates that a graphic design be selected based on the availability of an incremental algorithm. For this reason, simpler graphic designs may win out over ones that are prettier or more sophisticated.

2.3.2 *Radial Coordinates*

A very interesting visual effect is obtained by adopting a radial mapping in which execution sequence or time rotates around a point. Such mappings are used in a variety of visual metaphors. A radial mapping may represent similar information to that of a Cartesian mapping, but the user may recognize different patterns due to the metaphor employed. The center of the image provides a natural focus of attention for priority items such as the root of a tree. Figure 2.5 shows a circular tree with hundreds of nodes. More sophisticated radial techniques are possible with the introduction of hyperbolic geometry [12].

If 3D graphics are available, it is possible to obtain the best of both worlds. Radial coordinates may be used to lay out children around their parent at each level, while the third dimension allows the levels of the tree to be separated vertically as in traditional layouts. The result is attractive and scales well to handle large trees. One such layout is the cone tree [13]. Figure 2.6 shows an OpenGL depiction of a cone tree.

Some children may be hidden behind others. As long as the tool provides the ability to rotate subtrees at each node, any node can be brought to the front by some combination of rotations on its ancestors (Fig. 2.7).

2.3.3 Mapping Data to Colors, Textures, and Shapes

Visualizations often have many dimensions of information to depict. Horizontal and vertical coordinates provide two dimensions; animation may grudgingly provide a third. Color and texture provide fourth and fifth dimensions, with limitations.

An ubiquitous temptation in visualization is to use color to represent wide-ranging scalar values such as integers and floats. If you represent an entire numeric value in a single pixel, you have quite a bit better information density than writing out that number textually in a large rectangle of several hundred pixels. This has several limitations and problems, centering on the user's ability to interpret the color encoding correctly.

Unfortunately, the human viewer is not able to associate colors to numbers with a high degree of precision. Regardless of the display used, there is a tendency among untrained programmers writing visualizations to use the color spectrum to depict the range, mapping signed integers onto colors from violet to red. It turns out, though, that humans don't perceive the color spectrum in a linear fashion, so such a mapping is almost useless. Humans can be trained to learn it, but it has no inherent value.

Maps generally use a simpler range of colors to encode elevation using the range of colors one may see on the landscape at varying elevations, ranging from white snowy peaks on mountains to dark blue ocean depths. Such an encoding of numeric values is still quite artificial, but more likely to be usable than a spectrum-based coding. If you use a map-based color coding, at least those people familiar with the encoding will be able to read the numbers in your visualization.

Perhaps the best way of all to map numbers onto colors is to use grayscale. For example, the lowest possible value might correspond to black, while the highest value in the range is depicted as white. Humans are very good at discerning differences in the intensity of colors without changing the hue. For signed numbers, it is quite possible to use separate hues for positive and negative values. A zero value would be represented by a neutral black or white, and large values would progress to vivid red and blue, for example. This approach would be familiar to users of thermometers that depict red as hot and blue as cold.

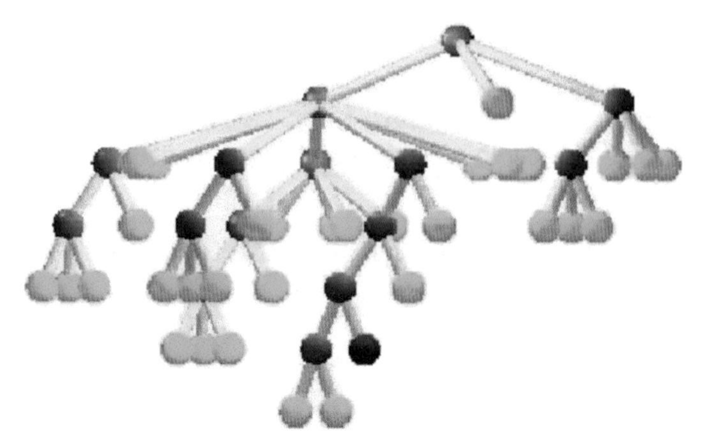

Fig. 2.6 A cone tree

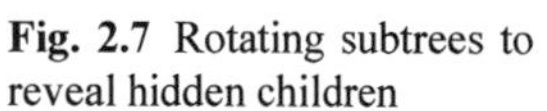

Fig. 2.7 Rotating subtrees to reveal hidden children

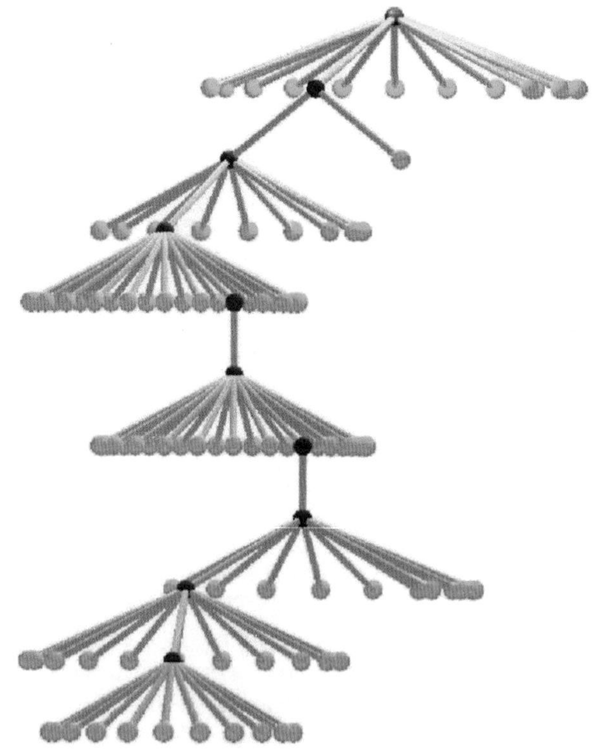

Textures are another means of depicting additional data within a given area. Solid area fills are generally better than textured fill patterns, so your first question when contemplating using texture is: can a color or grayscale substitute? Many people are partially or completely colorblind, so textures may be used in place of color to address that problem. The planets depicted in Fig. 2.4 show combinations of six different environment types using color and texture simultaneously; the fill patterns are reinforcing the color.

The eternal quest to render multidimensional data leads many programmers to try to depict features using textures and shapes in addition to color. This is usually a mistake, but if you are going to do it, you can at least note some of the following caveats. Any depiction that relies on textures or shapes is trading away resolution for dimensionality, just as dithering trades away resolution to simulate a larger number of colors.

With some exceptions, humans are not very good at memorizing the meanings of textures and shapes. You can get them to admit recognition of 7 ± 2 textures, such as "vertical stripes" or "crinkled paper," but that's not saying much. Humans can recognize lots more basic shapes such as diamonds or spades, but will still have trouble remembering the corresponding values for a large number of shapes. Also, shapes do not scale down; beneath a certain point they become illegible.

One of the most famous examples of depicting many dimensional data are Chernoff faces, where each multidimensional data point is represented by an entire face with separate values for eyes, ears, nose, and so on. This exploits humans' extensive face recognition training, which starts in early childhood. Humans may be pretty good at telling quickly if two faces are similar, but again each facial component has a small range of effective values; around seven is again a good guess.

References

1. J. Stasko, J. Domingue, M. Brown and B. Price, Eds., Software Visualization: Programming as a Multimedia Experience, Cambridge, MA: MIT Press, 1998.
2. E. Tufte, The Visual Display of Quantitative Information, Cheshire, CT: Graphics Press, 1983.
3. E. Tufte, Envisioning Information, Cheshire, CT: Graphics Press, 1990.
4. E. Tufte, Visual Explanations: Images and Quantities, Evidence and Narrative, Cheshire, CT: Graphics Press, 1997.
5. J. Bertin, Semiology of Graphics, Madison, Wisconsin: The University of Wisconsin Press, 1983.
6. D. Huff, How to Lie with Statistics, New York: Norton, 1954.
7. N. A. Salingaros, "Life and Complexity in Architecture from a Thermodynamic Analogy," *Physics Essays*, vol. 10, pp. 165–173, 1997.
8. S. Moen, "Drawing Dynamic Trees," *IEEE Software*, pp. 21–28, #jul# 1990.
9. B. Johnson and B. Schneiderman, "Tree-maps: A space-filling approach to the visualization of hierarchical information structures," in *IEEE Visualization '91 Conference Proceedings*, 1991.
10. G. Furnas, "Generalized Fisheye Views," in *CHI '86 Proceedings*, 1986.
11. M. Sarkar and M. H. Brown, "Graphical Fisheye Views," *Communications of the ACM*, vol. 37, pp. 73–84, 1994.
12. J. Lamping and R. Rao, "Laying out and Visualizing Large Trees Using a Hyperbolic Space," *Proceedings of ACM UIST '94*, vol. 18(3), no. 3, pp. 13–14, #nov# 1994.
13. G. G. Robertson, J. D. MacKinlay and S. K. Card, "Cone Trees: Animated 3D Visualizations of Hierarchical Information," in *Proceedings of CHI '91*, New Orleans, 1991.

Chapter 3
Software Instrumentation and Data Collection

The first rule of software visualization is: you cannot visualize what you cannot observe. Visualization depicts graphically many things that are not inherently graphical, but you can't visualize unless you can see inside the "black box" that is the thing you want to observe.

There are many pieces of electronic hardware that have been wired specially to be observable—they have hardware instrumentation. For example, an airplane is replete with instruments reporting on its state. In computing, the extent to which CPU makers have instrumented processors is very limited. Some CPU's do allow hardware monitoring, but even when they do, programmers need information at the source-code level; information at the machine-level is often too low-level.

For this reason, most software visualization starts with software instrumentation and data collection. In order for a visualization tool to report anything about some program X, the program is instrumented (possibly merely with print statements that generate a visualizable trace of the execution of various points in the program, but possibly something much more sophisticated).

This chapter covers:

- The Role of Static Information
- Log files versus runtime/real time event monitoring
- Events, and event monitoring, Unicon-style
- Inserting events in other languages' programs that can be reported and visualized in Unicon

3.1 The Role of Static Information

Statically determinable program properties, such as line counts or cyclomatic complexity numbers play a limited role in this book. Instead, the focus of this chapter is on visualizing dynamic behavior, because that is the more difficult task for which

C. Jeffery, J. Al-Gharaibeh, *Writing Virtual Environments for Software Visualization*, DOI 10.1007/978-1-4614-1755-2_3

visualization is more badly needed than understanding static information about the program. However, static information plays an important role: it provides context and aids in navigation through vast amounts of dynamic program behavior data. Static information helps in understanding dynamic information.

Note that "static" information from the point of view of a compiler writer is not really static, it is information that changes much more slowly than program execution information: on a human time scale instead of a CPU cycle time scale. If the human-scale change information is available via a version control system repository, it is useful to depict slow information changes in the static backdrop for dynamic visualizations, rather than pretend that the information is really static.

For the most part, then, the objective in the sections to follow will be to posit an aspect of program behavior, first ask how it can be observed, and then ask what static properties of the program provide a context for its interpretation. These three steps can be applied consistently to characterize applicable visual metaphors and tools for a given aspect of behavior.

3.2 Log Files and Real-Time Event Monitoring

Visualizing software happens either while it is running, or after it finishes. Visualizing while it is running interferes with its execution more, while visualizing afterwards requires creation of potentially gigantic log files, and constrains the ability to steer or ask what-if questions about the execution.

In this book we will generally favor real-time event monitoring, but a reasonable subset of execution monitors can be written to work either way. As long as a stream of program execution events provides all behavior information in question, a monitor might not care where that stream came from.

This chapter presents the Unicon program execution monitoring facilities. Unicon's monitoring facilities are a primary source of dynamic program execution information that is used in the virtual environment for software visualization described in this book. The monitoring facilities allow the user to execute a given Unicon program under the observation of one or more monitoring programs, also written in Unicon.

The overview begins with a brief inventory of the monitoring architecture and its components, followed by a user's eye view of a standard execution monitoring scenario. The purpose of the scenario is to characterize the execution monitoring process that is supported and to motivate some of the features and limitations of the system.

Following the execution monitoring scenario, the functional characteristics of each of the primary components of the execution monitoring framework are described. Details of the use of these components and their implementation are presented in subsequent chapters.

3.3 Inventory of Architecture Components

The monitoring facilities consist of the following components, summarized in Fig. 3.1. The monitoring components are characterized in terms of their relationship to related Unicon features. Many of these components are general-purpose language features that are useful independent of execution monitoring; such features, when already present in other languages, may require modification if they were not designed to support execution monitoring.

- The ability to load multiple programs into a shared execution environment is provided in order to adequately support monitor access to target program data. *Dynamic linking* is *not* desirable in the context of execution monitoring, since the names in the monitor are distinct from those in the target program.
- The monitor and target program execute independently, but not concurrently. This allows the monitor to control target program execution using a simple programming model. Unicon's *co-expression* data type supports synchronous execution of independent threads of execution; the mechanism is slightly extended to support the relationship between monitor and target program.
- Extensive information about program execution is available to the monitor from locations in the language runtime system that are coded to report significant events. This obviates the need for control-intrusive techniques of obtaining information from the target program. It also offers higher performance than target

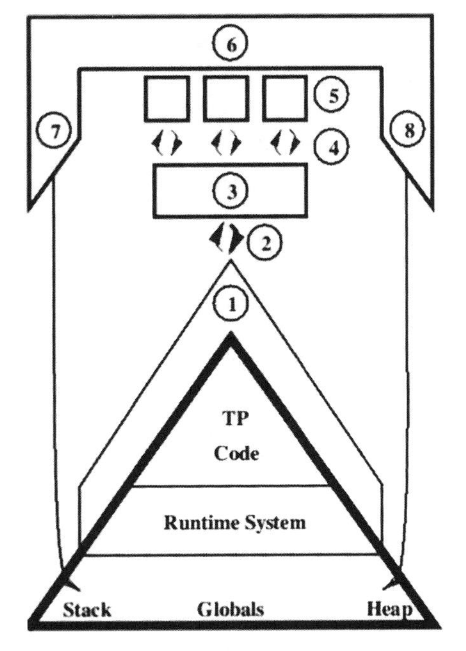

Fig. 3.1 Unicon's execution monitoring architecture

program instrumentation. The runtime system instrumentation is an extension and generalization of an earlier special-purpose monitoring facility oriented around dynamic memory allocation and reclamation [1]. It also supercedes the language's built-in procedure tracing mechanism [2].

- Monitor control over target program execution is coupled with the concept of *filtering* [3] in a mechanism called an *event mask*. Event masks provide a simple, dynamic model of execution control that adequately meets performance requirements in processing the high volume of execution information. Events that are of no interest to the execution monitor are never reported and do not impose unreasonable execution cost. Event masking uses a set abstraction to describe the execution behavior that is of interest to the monitor; an existing Unicon type that supports high performance set operations is employed to provide event masking in a manner that is familiar to Unicon programmers.

3.4 Standard Execution Monitoring Scenario

Understanding the monitoring architecture begins with a description of the activities that it supports. The following scenario presents the relationship between execution monitors and target program in its simplest form. More sophisticated relationships between the monitor and target program are discussed later in this chapter and in Chap. 12. In addition, the expected user and range of program behavior observable using these monitoring facilities are characterized.

target program (TP)—The target program is the Unicon program under study, a translated Unicon executable file. Monitoring does not require that the TP be recompiled, nor that the TP's source code be available, although some monitors make use of program text to present information.

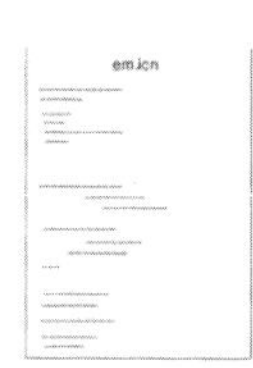

execution monitor (EM)—An execution monitor is a Unicon program that collects and presents information from an execution of a TP.

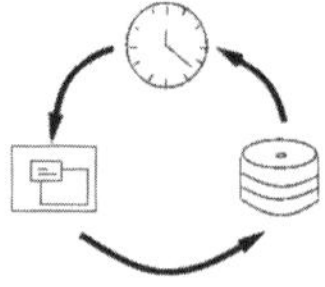

program behavior—Program behavior denotes the results of executing the TP. Behavior is meant in a general sense that includes program output, execution time, and the precise sequence of actions that take place during execution.

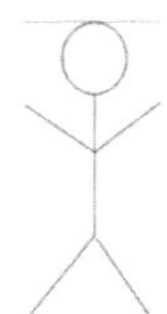

user—In our standard scenario, the user is a human capable of understanding the TP's execution behavior. The user must know the target language in order to make good use of many EMs or to write a new EM. In general, the user need not necessarily be familiar with the TP's source code.

3.4.1 Sources of Relevant Execution Behavior

Execution monitoring begins with a user who has questions about the behavior of a TP (Fig. 3.2). Typical questions relate to correctness or performance, such as "How is the result calculated?" or "What is taking so long?". Questions may be more general in nature if the user is trying to understand how a program works, rather than to change it.

Answers to important questions often may be found by following the execution as it proceeds through source language constructs, but in high-level languages the behavior in question often depends upon the language semantics as implemented by the language runtime system. Figure 3.3, iconx.c denotes the aggregate of files

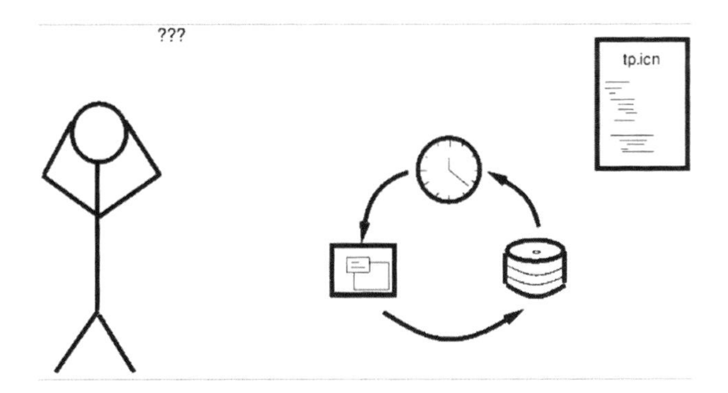

Fig. 3.2 Monitoring starts with a user, a program, and questions

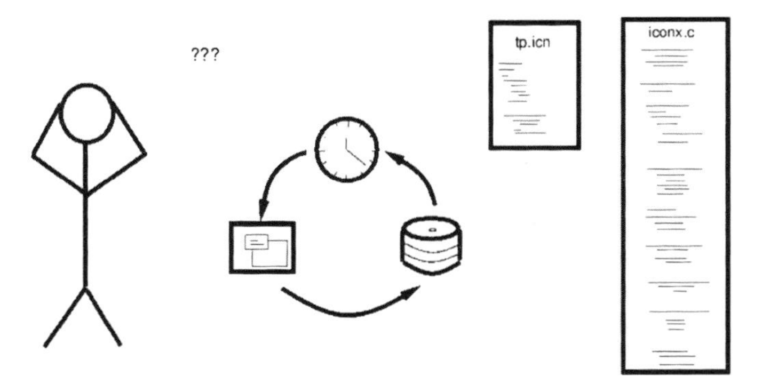

Fig. 3.3 Behavior depends on the language, not just the program

that comprise the Unicon language runtime system. For this reason, many forms of execution monitoring provide useful information even if the TP's source code is not available. Figure 3.3 could be further elaborated to include behavioral dependencies on the platform on which Unicon is implemented and run. Such dependencies are outside the scope of this book.

3.4.2 *Selecting or Developing Appropriate Monitors*

Rather than focusing on one monolithic EM that attempts to accommodate all monitoring tasks, the framework advocates development of a suite of specialized EMs that observe and present particular aspects of a TP's behavior. The user is responsible for selecting an appropriate EM or set of EMs that address the user's concerns.

If no available EM can provide the needed information, the user can modify an existing EM or write a new one. This end user development of execution monitors is also useful when an existing EM provides the needed information, but it is obscured by other information; existing EMs can be customized to a particular problem.

3.4.3 *Running the Target Program*

The user runs the TP one or more times, monitored by a selection of EMs (Fig. 3.4). General-purpose EMs provide an overall impression of program behavior. Visualization techniques enable the presentation of a large amount of information and abstract away detail.

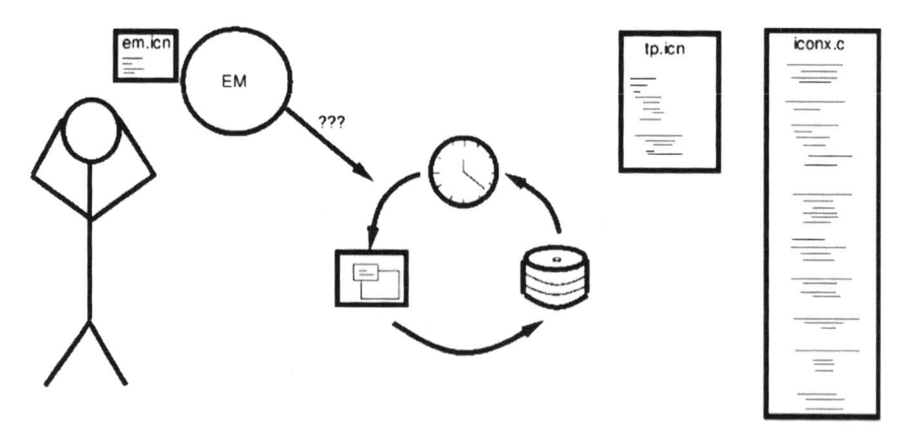

Fig. 3.4 EMs can answer questions about TP behavior

Obtaining more specific information frequently requires that the user interact with the EMs to control the TP's execution, either to increase the amount of information presented during specific portions of execution or to stop execution in order to examine details. In order to provide this interactive control, EMs must present execution information as it happens during the TP's execution, rather than during a postmortem analysis phase.

3.5 Framework Characteristics

The preceding scenario depends on support for exploratory programming in several areas: controlling a program's execution, obtaining execution information, presenting large quantities of information, and interacting with the user. In order to support these tasks, the framework provides synchronous shared address multitasking and an event-driven execution control model. These features are provided by extensions to the Unicon language.

3.5.1 Multitasking

The first and most basic characteristic of the framework is an execution model in which an EM is a separate program from the TP—a multitasking model. In this model the EM views the TP as a separately loaded coroutine [4]. The coroutine relationship is the primary means by which EMs control TP execution, and coroutine transfers of control are the primary source of execution information from a TP (Fig. 3.5). The precise nature of the interaction between the EM and TP (the arrows in Fig. 3.5) is discussed in the next section.

Multitasking has the following benefits in an exploratory programming environment: the EM and TP are independent programs, the EM has full access to the TP, and the mechanism accommodates multiple EMs. These benefits are described in more detail below.

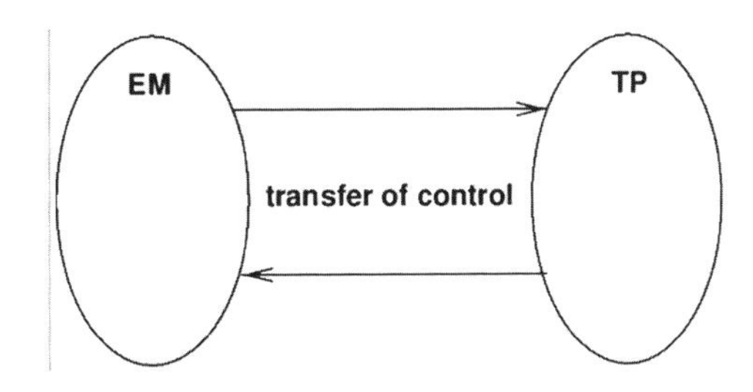

Fig. 3.5 EM and TP are separately loaded coroutines

3.5.2 Independence

Because the EM and TP are separate programs, the TP need not be modified or even recompiled in order to be monitored by an EM; neither does an EM need modification or recompilation in order to be used on different target programs. The separation of EMs and TPs also simplifies the writing of EMs because an EM need not be implemented as a set of callback functions—it has its own control flow. By definition, execution of tasks such as EMs and TPs is synchronous. The TP is not running when an EM is running, and vice versa. This synchronous execution allows EMs and TPs to be independent without introducing the complexity inherent in concurrent programming.

Another degree of EM and TP independence is afforded by separate memory regions; EMs and TPs allocate memory from separate heaps. For this reason memory allocation in the EM does not affect the allocation and garbage collection patterns in the TP. Because Icon is a type-safe language with runtime type checking and no pointer data types, EMs and TPs cannot corrupt each others' memory by accident; only code that contains explicit references to another program's variables and data can modify that program's behavior. EMs can (and some do) modify TP values in arbitrary ways; the purpose of separate memory regions is to minimize *unintentional* data intrusion.

3.5.3 Access

An address space is a mapping from machine addresses to computer memory. Within an address space, access to program variables and data is direct, efficient operations such as single machine instructions. Accessing program variables and data from outside the address space is slower and requires operating system assistance.

In Unicon, programs such as the EM and TP reside within the same address space. This allows EMs to treat TP data values in the same way as their own: EMs can access TP structures using regular Icon operations, compare TP strings with their own, and so forth.

Because of the shared address space, the task-switching operation needed to transfer execution between EMs and TPs is a fast, lightweight operation. This is important because monitoring requires an extremely large number of task switches compared to typical multitasking applications.

3.5.4 Multiple Monitors and Monitor Coordinators

Unicon's dynamic loading capabilities allow simultaneous execution of not just a single EM and a single TP, but potentially many EMs, TPs, and other Icon programs in arbitrary configurations. Although uses for many such configurations can

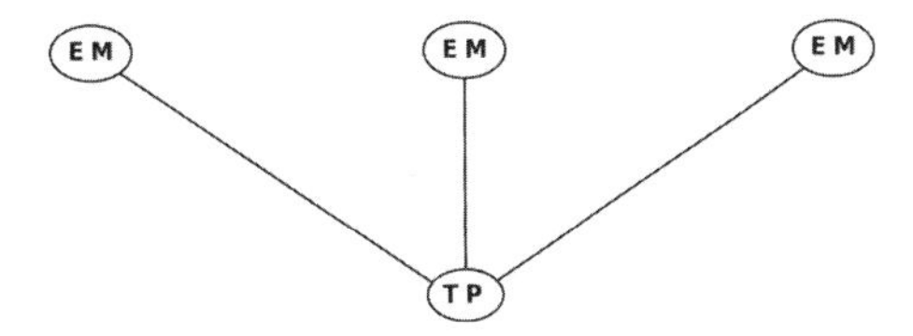

Fig. 3.6 Multiple EMs

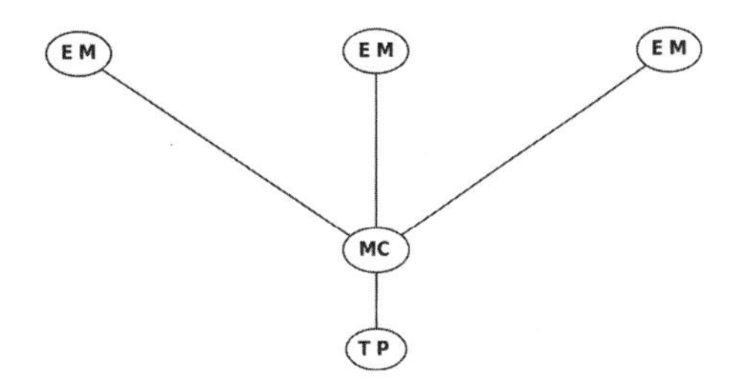

Fig. 3.7 An execution monitor coordinator

be found, one configuration merits special attention when many specialized EMs are available: the execution of multiple monitors on a single TP (Fig. 3.6).

The difficulty posed by multiple monitors is not in loading the programs, but in coordinating and transferring control among several EMs and providing each EM with the TP execution information it requires. Since EMs are easier to write if they need not be aware of each other, this motivates construction of *monitor coordinators* (MCs), special EMs that monitor a TP and provide monitoring services to one or more additional EMs (Fig. 3.7). EMs receiving an MC's services need not be aware of the presence of an MC any more than a TP need be aware of the presence of an EM.

The virtual monitor interface provided by MCs makes adding a new monitor to the system extremely easy. A new monitor could conceivably be written, compiled, linked, and loaded during a pause in the TP's execution. In addition, constructing new MCs that provide high-level services is a straightforward task in Unicon and is how monitors are embedded within virtual environments in this book: the virtual environment is a monitor coordinator that allows monitors to observe the target program and embed their graphics within a surrounding 3D context.

3.5.5 Execution Control

The primary task of an EM is to collect data from a TP's execution. This task poses difficult coding problems and is frequently a performance bottleneck. The nature of the data collection facilities available in a monitoring system also define and limit the kinds of monitors that can be implemented.

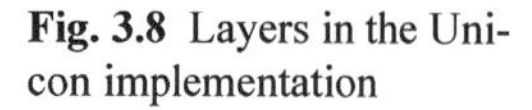

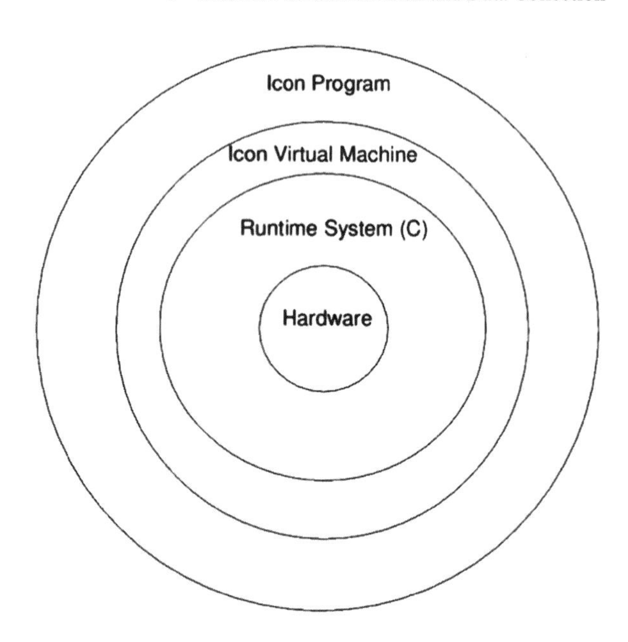

Fig. 3.8 Layers in the Unicon implementation

Figure 3.8 depicts the system layers present in running a program under the Unicon interpreter. The TP code is executed by a virtual machine interpreter written in C, which in turn calls C language runtime support code to perform various language operations [5].

Of these layers, the TP code, the virtual machine (VM), and the runtime support code are responsible for aspects of program behavior within the scope of this research. The VM and the runtime system have been extensively instrumented to produce this information for EMs at the Unicon level without requiring instrumentation of the TP code.

While the behavior observable from instrumentation of the VM is specific to the Unicon interpreter and is of interest primarily to language implementors, runtime system behavior is more general and of interest to normal programmers. This book is primarily concerned with monitors of runtime system behavior. Most of this behavior takes place even in compiled native-code versions of the TP, with the exception of behavioral aspects such as runtime type checks that a compiler can avoid when static analysis determines that they are unnecessary.

This instrumentation consists of locations within the runtime system at which control can be transferred and information reported to the EM. When execution proceeds through one of these points in the runtime system, an *event* occurs. Many events take place during even the simplest of Icon operations. When an EM resumes execution of the TP, it explicitly specifies what kinds of events are to be reported; other kinds of events are not reported. The kinds of events to be reported can be changed dynamically each time the TP's execution is resumed (Fig. 3.9). The processing of an event includes a test of whether the TP should transfer control to the EM and code to perform the transfer only if the test succeeds.

Those events at which control is transferred produce *event reports*. When an event is reported, the TP's execution is suspended and execution commences in

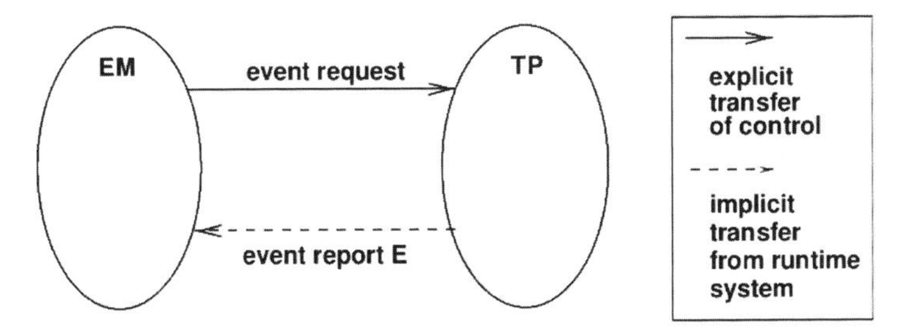

Fig. 3.9 Event-driven control of TP

the program that loaded the TP—an EM. Event reporting supports data collection in two ways: an event report contains some information associated with the event itself, and in addition, when the EM gains control, it can interrogate the TP's variables and keywords for further information. When an EM requests another event report, the EM suspends execution and the TP's execution resumes where it left off.

3.5.6 Visualization Support

Unicon's monitoring facilities explicitly support graphics output and user interaction in EMs. Given the amount of information associated with the execution of TP, most EMs use graphical techniques to present abstractions of execution information. Since the monitor cannot in general anticipate what information will be relevant or how to interpret it, user interaction is crucial to the success of the monitoring process.

3.5.7 Simplified Graphics Programming

Unicon includes a high-level interface to computers' graphical display facilities; the language provides a built-in window type. The window and any associated window system parameters such as the graphics drawing context and display connection are implicitly bound together as a single value. Programs with a primary window can designate it as the implicit subject of all window operations.

Graphic displays and window system software contain a variety of resources such as colors, fonts, and images. These resources are allocated implicitly by the system when they are used and require minimal attention by the user.

3.5.8 Support for Dual Input Streams

An EM typically has two primary input streams: the event stream from the TP, and the input stream from the user (Fig. 3.10). Although these two input streams are conceptually independent and may be treated as such, for many EMs this

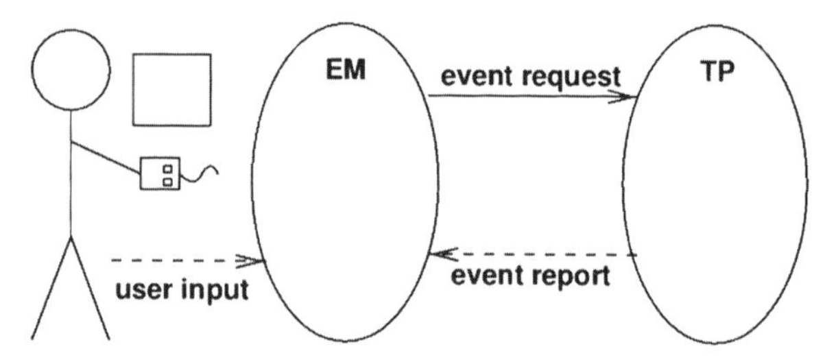

Fig. 3.10 Monitors have two input streams

unnecessarily complicates the central loop that obtains event reports from TP—the EM must also check its own window for user activity.

The runtime system instrumentation includes code that optionally checks for EM input and reports it as an event by the execution monitoring facility, instead of requiring that the EM explicitly check the user input stream. This simplifies EM control flow and improves EM performance.

3.5.9 Inter-task Access Functions

Several built-in functions provide inter-task access to program data. For example, the `variable()` function in Unicon takes a co-expression value as an optional second argument denoting the task from which to fetch the named variable. When called with this second argument, `variable()` is useful for assigning to or simply reading values from another task's variables. In this example, the parent task initializes each child task's `parent` global variable (if there is one) to refer to the parent's `&main` co-expression. A child task can then use this variable to determine whether it is being run standalone or under a parent task. Inter-program access through the `variable()` function is also useful in inspecting values, especially at intermediate points during the monitored execution of a TP, as described in the next chapter.

```
procedure main(arguments)
  every arg := !arguments do {
    Task := load(arg)
    variable("parent", Task) := &main
    @Task
    }
```

In addition to Unicon's extensions of existing functions, several new functions have been added. These facilities are useful in execution monitoring and are used in examples later in this book. Some of the intertask access functions used in examples are listed in Table 3.1. In these functions, parameter C refers to a co-expression that may be from a task other than the one being executed. Functions that are said to *generate* values can produce more than one result from a given call.

Table 3.1 Unicon interprogram access functions

`cofail(C)`	Transmit failure to `C`
`fieldnames(r)`	Generate fieldnames of record `r`
`globalnames(C)`	Generate the names of `C`'s global variables
`keyword(s, C)`	Produce keyword `s` in `C`. Keywords are special global variables that have special semantics in certain language facilities
`localnames(C,i)`	Generate the names of `C`'s local variables, `i` calls up from the current procedure call
`paramnames(C,i)`	Generate the names of `C`'s parameters
`staticnames(C, i)`	Generate the names of `C`'s static variables
`structure(C)`	Generate the values in `C`'s block region, or heap. The heap holds structure types such as lists and tables
`variable(s,C,i)`	Produce variable named `s`, interpreted `i` levels up within `C`'s procedure stack

3.6 Events

The primary linguistic concept added in order to support execution monitoring is an *event*. An event is the smallest unit of execution behavior that is observable by a monitor. In practice, an event is the execution of a point of instrumentation in the code called a *sensor* [6] that is capable of transfering control to the monitor.

This definition limits events to those aspects of program behavior that are instrumented in the language runtime system or the program itself. The event model is only as useful or general as is the instrumentation that extracts program information. If instrumentation does not exist for an aspect of program behavior of interest, it is often possible to monitor the desired behavior by means of other events. In the present implementation, for example, no instrumentation exists for file input and output. If an EM wishes to monitor I/O behavior, it can monitor function and operator events and act on those functions and operators that relate to input and output.

The Unicon definition of event also differs from that of many monitoring systems, in which the term event refers to the basic unit of information *received* by the monitor [7]. The distinction is that in the Unicon definition, events occur whether they are monitored or not, and each event may or may not be observed by any particular monitor. This definition is useful in the Unicon environment, in which EMs are not coupled with the instrumentation and multiple EMs can observe a TP's execution.

3.6.1 Event Codes and Values

From the monitor's perspective, an event has two components: an *event code* and an *event value*. The code is generally a one-character string describing what type of event has taken place. For example, the event code C denotes a procedure call event. Event codes all have associated symbolic constants used in program source code. For example the mnemonic for a procedure call event is `E_Pcall`. These constants are available to programmers as part of a standard event monitoring library described below.

The event value is an Unicon value associated with the event. The nature of an event value depends on the corresponding event code. For example, the event value for a procedure call event is an Unicon value designating the procedure being called, the event value for a list creation event is the list that was created, the event value for a source location change event is the new source location, and so forth. Event values can be arbitrary Unicon structures with pointer semantics; the EM accesses them just like any other source language value.

3.6.2 Event Reporting and Masking

The number of events that occurs during a program execution is extremely large—large enough to create serious performance problems in an interactive system. Most EMs function effectively on a small fraction of the available events; the events that an EM uses are said to be *reported* to the EM. An *event report* results in a transfer of control from the TP to the EM. Efficient support for the selection of appropriate events to report and the minimization of the number of event reports are primary concerns.

Unicon supports *dynamic event masking* based on event codes, a dynamic variation of the filter concept found in most event-based monitoring systems [3, 7]. Event masking allows the monitor to specify which events are to be reported and to change the specification at runtime. When the program being monitored starts execution, the monitor selects a subset of possible event codes from which to receive its first report. The program executes until an event occurs with a selected code, at which time the event is reported. After the monitor has finished processing the report, it transfers control back to the program, again specifying an event mask. Dynamic event masking enables the monitor to change the event mask in between event reports.

The use of one character strings as event codes has a more practical value than its mnemonic merit: it allows sets of codes to be efficiently and easily manipulated at the Unicon level by the *cset* (character set) data type. Csets are represented internally by bit vectors, so a cset membership test is very efficient compared to Unicon's more generic set data type, whose membership test is a hash table lookup.

When an event report transfers control from TP to EM, the two components of the event are supplied in the keywords `&eventcode` and `&eventvalue`, respectively. As discussed in the preceding chapter, these keywords are special global variables that are given their values by the Unicon runtime system during an event report, rather than by explicit user assignment. The monitor then can act upon the event based on its code, display or manipulate its value, etc.

3.6.3 *Obtaining Events Using evinit*

A standard library called `evinit` provides EMs with a means of obtaining events. Programs wishing to use the standard library include a link declaration such as `link evinit`. In addition, monitors include a header file named `evdefs.icn` to obtain the symbolic names of the event codes.

3.6.4 *Setting Up an Event Stream*

An EM first sets up a source of events; the act of monitoring then consists of a loop that requests and processes events from the TP. Execution monitoring is initialized by the procedure `EvInit(x [,input, output,error])`. If `x` is a string, it is used as an icode file name in a call to the Unicon function `load()`. If `x` is a list, its first argument is taken as the icode file name and the rest of the list is passed in to the loaded function as the arguments to its main procedure. `EvInit()` assigns the loaded TP's co-expression value to EM's `&eventsource` keyword. The input, output, and error arguments are files used as the loaded program's standard files.

EMs generally call the library procedure `EvTerm()` when they complete, passing it their main window (if they use one) as a parameter. `EvTerm()` informs the user that execution has completed and allows the final screen image to be viewed at the user's leisure, waiting for the user to press a key or mouse button in the window and then closing it.

The typical EM, and all of the EMs presented as examples in this book, follow the general outline:

```
$include "evdefs.icn"
link evinit
procedure main(arguments)
   EvInit(arguments) | stop("can't initialize monitor")
   # ... initialization code, open the EM window
   # ... event processing loop (described below)
   EvTerm()
end
```

This template is generally omitted from program examples for the sake of brevity.

3.6.5 *EvGet()*

Events are requested by an EM using the function `EvGet(mask)`. `EvGet(mask)` activates the co-expression value of the keyword `&eventsource` to obtain an event whose code is a member of the cset `mask`. `mask` defaults to `&cset`, the universal set indicating all events are to be reported. The TP executes until an event report takes place; the resulting code and value are assigned to the keywords `&eventcode` and `&eventvalue`. `EvGet()` fails when execution terminates in TP.

`EvGet()` allows a monitor the flexibility to change event masks each time the event source is activated. Another function that sets event masks is `eventmask()`. `eventmask(C,c)` sets the event mask of the task owning co-expression `C` to the cset value given in `c`.

3.6.6 Event Masks, and Value Masks

Event masks are the primary filtering mechanism in Unicon's monitoring facilities, but there are situations where they are not specific enough. For example, instead of handling events for all list operations, you may want events only for specific lists. This situation is supported by the concept of *value masks*. A value mask is a set or cset whose members are used to filter events based on their `&eventvalue`, just as an event mask filters based on the `&eventcode`. You may specify a different value mask for each event code. Value masks for all event codes are supplied in a single table value whose keys map event codes to corresponding value masks. This table is passed as an optional second parameter to `EvGet()` or third parameter to `eventmask()`. Note that no value mask filtering is performed for event codes that are not key in the value mask. Note also that value masks persist across calls to `EvGet()`. They are replaced when a new value mask is supplied, or disabled if a non-table is passed as the value mask parameter.

There is one special case of value masks that receives extra support: virtual machine instructions. Requesting an event report for the execution of the next virtual machine instruction is performed by calling `EvGet()` with an event mask containing `E_Opcode`. VM instructions occur extremely frequently; dozens of them can occur as a result of the execution of a single line of source code. Consequently, performance is severely affected by the selection of all VM instruction events.

However, a particular instruction or small set of instructions may be of interest to a monitor. In that case, the EM need not receive reports for all instructions. The function `opmask(C,c)` allows an EM to select a subset of virtual machine instructions given by `c` in `c`'s task. Subsequent calls to `EvGet()` in which `E_Opcode` is selected reports events only for the VM instructions designated by `c`.

The event values for `E_Opcode` are small non-negative integers. They fall in a limited range (<256), which is what allows a cset representation for them. Symbolic names for individual virtual machine instructions are defined in the include file `opdefs.icn`. `opmask(C,c)` is equivalent to:

```
t[E_Opcode] := c
eventmask(C, , t)
```

3.6.7 Artificial Events

As described above, the Unicon co-expression model allows interprogram communication via explicit co-expression activation or implicit event reporting within the runtime system. *Artificial events* are events produced by explicit Unicon code; they can be viewed at the language level as co-expression activations that follow the same protocol as implicit events, assigning to the keyword variables `&eventcode` and `&eventvalue` in the co-expression being activated.

There are two general categories of artificial events, *virtual events* meant to be indistinguishable from implicit events and *pseudo events* that convey control messages to an EM. Virtual events are generally used either to produce event reports from manually instrumented locations in the source program, to simulate event reports, or to pass on a real event from the primary EM that received it to one or more secondary EMs. Pseudo events, on the other hand, are used for more general inter-tool communications during the course of monitoring, independent of the TP's execution behavior. The function `event(code, value, recipient)` sends a virtual event report to the co-expression `recipient`, which defaults to the `&main` co-expression in the parent of the current task, the same destination to which implicit events are reported.

There are times when a primary EM wants to pass on its events to a secondary EM. An example would be an event transducer that sits in between the EM and TP, and uses its own logic to determine which events are reported to EM with more precision than is provided by the masking mechanism. A transducer might just as easily report extra events with additional information it computes, in addition to those received from TP. A more substantial application of virtual events is a monitor coordinator, an EM that coordinates and produces events for other monitors. Monitor coordinators are presented in Chap. 5.

EMs generally have an event processing loop as their central control flow mechanism. The logical way to communicate with such a tool is to send it an event. In order to distinguish a message from a regular event report, the event code must be different from those of regular events. In the monitoring framework, this is achieved simply by using an event code other than a one-letter string, such as an integer. Since not all EMs handle such events, they are not delivered to an EM unless it passes a non-null second argument (the "value mask argument") to `EvGet()`, such as `EvGet(mask,1)`.

The framework defines a minimal set of standard pseudo events, which well-behaved EMs should handle correctly; these pseudo events are described in Chap. 5. Beyond this minimal set, pseudo events allow the execution monitor writer to explore communication between EMs as another facility to ease programming tasks within the monitoring framework.

3.7 Instrumentation in the Unicon Runtime System

This section describes the instrumentation used by Unicon to produce events at various points in the runtime system. Significant points in interpreter execution where transfer of control might be warranted are explicitly coded into the runtime system with tests that result in transfer of control to an EM when they succeed. When execution reaches one of these points, an event occurs. Events affect the execution time of the TP; execution is either slowed by a test and branch instruction (if the event is not of interest to the EM), or stopped while the event is reported to the EM and it processes information. Minimizing the slowdown incurred due to the presence of monitoring instrumentation has been a focus of the implementation.

There are several major classes of events that have been instrumented in the Unicon intepreter. Most of these events correspond to explicit elements within the source code; others designate actions performed implicitly by the runtime system that the programmer may be unaware of. A third class of event that has been instrumented supports user interaction with the EM rather than TP behavior.

3.7.1 Explicit Source-Related Events

The events that relate behavior observable from the source code are:

- Source code locations are reported in terms of line numbers and columns.
- There are events for procedure calls, returns, failures, suspensions, and resumptions. In addition to these explicit forms of procedure activity, events occur for implicit removals of procedure frames.
- Events that correspond to Unicon built-ins describe many areas of behavior from numeric and string operations to structure accesses and assignments. Like procedures, events are produced for function and operator calls, returns, suspensions, resumptions, and removals.
- Unicon's pattern matching operations include scanning environment creation, entry, change in position, and exit. To obtain a complete picture of string scanning, monitors must observe these events along with the built-in functions related to string scanning.

3.7.2 Implicit Runtime System Events

Events that depict important program behavior observed within the runtime system include:

- Memory is allocated from the string and block regions in the heap. Allocation events include size and type information. This instrumentation is based on earlier instrumentation added to Icon for a memory monitoring and visualization system [1].

- The storage region being collected (Unicon has separate regions for strings and data structures), the memory layout after compaction, and the completion of garbage collection are reported by several events.
- In Unicon, automatic conversions are performed on parameters to functions and operators. Information is available for conversions attempted, failed, succeeded, and found to be unnecessary.
- Unicon's semantics may be defined by a sequence of instructions executed by the Unicon virtual machine [5]. The program can receive events for all virtual machine instructions, or an arbitrary subset.
- The passage of CPU time is indicated by a clock tick.

Most EMs, except completely passive visualizations and profiling tools, provide the user with some degree of control over the monitoring activity and must take user interaction into account. For example, the amount of detail or the rate at which the monitor information is updated may be variables under user control. Since an EM's user input occurs only as often as the user presses keys or moves the mouse, user interaction is typically far less frequent than events in the TP. Even if no user input occurs, polling for user input may impose a significant overhead on the EM because it adds code to the central event processing loop.

In order to avoid this overhead, the event monitoring instrumentation includes support for reporting user activity in the EM window as part of the TP's event stream. Monitor interaction events are requested by the event code `E_MXevent`. An example of the use of monitor interaction events is presented further in this chapter in Sect. 3.8.1, entitled "Handling User Input". A complete list of event codes is presented in Appendix A in order to indicate the extent of the instrumentation.

3.8 Anatomy of an Execution Monitor

The execution monitoring interface presented in this chapter uses a form of *event driven* programming: the central control flow of EM is a loop that executes the TP for some amount of time and then returns control to EM with information in the form of an event report. The central loop of an EM typically looks like:

```
case &eventcode of {
  # a case clause for each code in the event mask
  }
```

Event-driven programming is more commonly found in programs that employ a graphical user interface, where user activity dominates control flow. Because monitoring employs a programming paradigm that has been heavily studied, many coding techniques developed for graphical user interface programming, such as the use of callbacks [8], are applicable to monitors. Several of the example EMs in subsequent chapters use a callback model to take advantage of a higher-level monitoring abstraction available by means of a library procedure.

3.8.1 *Handling User Input*

An EM that handles user input could do so by polling the window system after each event in the main loop:

```
case &eventcode of {
  # a case clause for each code in the event mask
  }
# poll the window system for user input
```

If the events being requested from the TP are relatively infrequent, this causes no great problem. However, the more frequent the event reports are, the more overhead is incurred by this approach relative to the execution in TP. In typical EMs, polling for user events may slow execution from an imperceptible amount to as much as 15%.

Since the slowdown is a function of the frequency of the event reports and not just the cost of the polling operation itself, techniques such as maintaining a counter and polling once every *n* event reports still impose a significant overhead. In addition, such techniques reduce the responsiveness of the tool to user input and therefore reduce the user's control over execution.

Monitor interaction events, presented earlier in this chapter, address this performance issue by allowing user input to be supplied via the standard event stream produced by `EvGet()`. Since the `E_MXevent` event occurs far less frequently than other events, it makes sense to place it last in the case expression that is used to select actions based on the event code. Using this feature, the main loop becomes:

```
case &eventcode of {
  # other cases update image to reflect the event
  E_MXevent: {
    # process user input event
    }
  }
```

`EvGet()` reports pending user activity immediately when it is available; the control over execution it provides is comparable to polling for user input on each event.

3.8.2 *Querying the Target Program for more Information*

After each event report, EMs can use Unicon's intertask data access functions to query TP for additional information, such as the values of program variables and keywords. The access functions can be used in several ways, such as

- Applying a predicate to each event report to make monitoring more specific,
- *Sampling* execution behavior not reported by events by polling the TP for information unrelated to the event reports [6], or
- Presenting detailed information to the user, such as the contents of variables.

3.8.3 2D Graphics Capabilities

Unicon is known as a string and list processing language, but it also includes extensive graphics facilities. Visualization tools written in Unicon present their output using the type window. This section describes only the graphics functions that are used in subsequent chapters. See the book by Griswold, Jeffery, and Townsend for an extensive description of Unicon's 2D graphics facilities [9].

Windows allow both text and graphic input/output to be freely mixed. While on screen, windows may be moved, resized, and iconified by the user or the Unicon program. Window exposure (also known as redraw or paint) events are handled automatically and do not have to be handled by the programmer; the window contents are *retained* until the window closes. If the keyword `&window` has a window value, it serves as a default window for all graphic functions. The remaining examples in this chapter assume `&window` is the window of interest.

Unicon's window interface uses a raster graphics model. In this model, a window is a two-dimensional array of points, also called picture elements (*pixels*) in the x- and y-coordinates starting from the pixel (0,0) in the upper-left corner and moving positive to the right and down the window. Several functions take pixel coordinates and draw geometric figures on the window. Pixels are drawn with a window's current *foreground color*. Some useful functions are given in Table 3.2; other graphics functions are described as they are used in examples.

Many visualization tools make extensive use of color in graphics operations to encode information about related data types or program operations. Such tools could change the output drawing color by repeated calls to `Fg()`, but it is much faster to ask the window system to set up several window values that draw with different colors. The call `Clone(&window,"fg="||s)` creates a window value that draws on `&window` using foreground color `s`. All graphics functions

Table 3.2 Some useful graphics functions

`EraseArea()`	Clears a rectangular area
`DrawArc()`	Draws an arc
`DrawPoint()`	Draws a point
`DrawLine()`	Draws a line
`DrawRectangle()`	Draws a rectangle
`DrawString()`	Draws a string
`Event()`	Returns the next user event
`Fg()`	Sets the color used in subsequent drawing
`FillArc()`	Draws a filled arc
`FillRectangle()`	Draws a filled rectangle
`GotoRC()`	Moves text cursor position
`Pending()`	Returns a list with user events awaiting processing
`WAttrib(w, attr)`	Get/set a window attribute such as height or width

may be prefixed with such a window argument w to draw with a nondefault color, for example

```
DrawPoint(w_red, x, y)
```

draws a red point at (x, y).

When an encoding of colors is used in a visualization tool, a table is typically used to store a mapping from a source domain such as string type names to window bindings with various colors.

3.8.4 Some Useful Library Procedures

Several library procedures are useful in EMs. This section presents those library procedures that are included in the evinit library. Location decoding and encoding procedures are useful in processing location change event values, but they are also useful in other monitors in which two-dimensional screen coordinates must be manipulated. Besides program text line and columns, the technique can variously be applied to individual pixels, to screen line and columns, or to screen grid locations in other application-specific units. In addition, various EMs use utility procedures. Table 3.3 lists the library procedures that are used in this book.

Table 3.3 Additional library procedures for monitors

Procedure	Returns or computes
evnames(s)	Converts event codes to text descriptions and vice versa
evsyms()	Two-way table mapping event codes to their names
typebind(w, c)	Table mapping codes to color-coded clones for w
opnames()	Table mapping VM instructions to their names
location()	Encodes a two dimensional location in an integer
vertical()	Y/line/row component of a location
horizontal()	X/column component of a location
prog_len()	Number of lines in the source code for TP
procedure_name()	Name of a procedure
Wcolumns()	Window width in text columns
Wheight()	Window height in pixels
Wrows()	Window height in text rows
Wwidth()	Window width in pixels

3.9 Typical Evolution of a Visualization Tool

Many visualization tools can be written by adapting a similar existing tool; in fact, many of the simple visualizations described later in this book are intended as starting points for such programs. Because program visualization is still a relatively undeveloped area, however, there are still frequent situations where a new monitor is written to visualize a new behavior of interest. In such a situation, the monitor writer gets to work from scratch.

While every programmer may approach the task of writing a visualization tool differently, we found over time that a consistent approach has been useful for a wide range of tools. Our method is incremental and reflects the view that the monitoring task is predominant over the graphics programming task in constructing such tools. We present the sequence of tasks here in order to encapsulate this experience. Following the approach is no guarantee that the end product will be successful, but it may simplify for the reader the order in which various components are best assembled.

3.9.1 Generate Log Files

The initial emphasis should be to characterize precisely the behavior of interest. Program execution behavior is expressed in terms of a stream of events; example code for handling them was given earlier in this chapter. A prose description of the behavior of interest may be useful, but a state machine or grammar that describes the event sequences that constitute the behavior of interest is a more useful formalization.

Some behavior will not be described adequately by a regular or context-free sequence of events. In particular, some monitors may have to examine variables and check complex conditions within the program being observed, in order to find the behavior of interest. For these reasons, the ultimate formal preparation prior to writing the visualization code is to produce a fully operational monitor that observes the desired event sequences and writes the relevant events to a logfile. This text-only monitor can be tested to establish confidence in it before starting the task of graphics coding.

3.9.2 Depict the Log Files

Once a monitor that observes the correct behavior is written, the primary graphic design and programming can be performed. In the absence of an obvious metaphor, the same information that was being written to the logfiles can be animated as a sequence of points or more complex objects plotted along a Cartesian or radial axis.

3.9.3 Scale to Handle Real Problems

For most monitors, multiple forms of scaling are necessary in order to produce a usable tool from the initial graphic design. Traditional scaling involves multiplying coordinates by a scaling factor in order to ensure that the plotted coordinates do not go outside the visible area of the window. Other forms of scaling begin the conversion from graphs to visualizations. This includes condensation of multiple events down to individual output operations, and of multiple graphic outputs onto a shared space in the window. Several iterations of scaling may be applied as needed.

3.9.4 Focus on Behaviors of Interest

Scaling often leads to enough loss of detail that the user cannot see the behaviors of interest, they can only see the big picture. For cataclysmic events of interest, the monitor can help the user notice behaviors of interest by drawing attention to them using flashing lights, sound, or whatever. For more nebulous or tenuous situations, the monitor's goal should be to show enough information for users to decide whether something needs further investigation. Showing more details for atypical behavior is a good starting point. To show more details, the tool requires more screen space. Log scales or hyperbolic or fisheye views allocate conspicuously more screen space to some elements than to others.

3.9.5 Add User-Directed Navigation

Ultimately, the user will need to be able to specify objects on the visualization for which additional detailed information is of interest. Adding navigation may be more difficult than rendering the graphics in the first place. The visualization tool will need to perform hit tests to determine which object the user is selecting. This requires a data structure that provides access to on-screen entities, and efficiency may become an issue for very large numbers of objects. This task typically leads to the additional graphics programming required to produce detailed views of objects of interest.

3.10 Instrumenting Code Written in Other Languages

All this talk about monitoring Unicon programs is great, but what about programs written in other, more mainstream languages? This section describes how to insert events in other languages' programs that can be reported and visualized in Unicon using the tools presented in this book.

Any programming language whose program execution behavior can be abstracted by a mapping onto a Unicon event stream can relatively easily be visualized using the virtual environment and tools described in this book. The main issue will be whether performance is sufficient for real-time monitoring while the program is running, or whether monitoring had better be done into a log file, and visualization performed *post-mortem*.

Log files run very quickly into gigabyte territory. There was a time when run-time monitoring was more attractive specifically because it was desirable to avoid having to store enormous log files. In the era of multi-terabyte PC hard drives, space is less of a concern. Real-time monitoring still retains the advantage, potentially, of interacting with the execution being monitored with the possibility of altering, or steering the execution.

References

1. R. E. Griswold and G. M. Townsend, “The Visualization of Dynamic Memory Management in the Icon Programming Language,” 1989.
2. R. E. Griswold and M. T. Griswold, The Icon Programming Language, third edition, San Jose, California: Peer-to-Peer Communications, 1997.
3. I. J. P. Elshoff, “A Distributed Debugger for Amoeba,” in Proceedings of the ACM SIGPLAN/ SIGOPS Workshop on Parallel and Distributed Debugging ACM SIGPLAN Notices /, 1989.
4. C. Marlin, Coroutines (Lecture Notes in Computer Science 95), Berlin: Springer-Verlag, 1980.
5. R. E. Griswold and M. T. Griswold, The Implementation of the Icon Programming Language, Princeton, New Jersey: Princeton University Press, 1986.
6. D. M. Ogle, K. Schwan and R. Snodgrass, “The Dynamic Monitoring of Distributed and Parallel Systems,” 1990.
7. P. Bates, “Debugging Heterogeneous Distributed Systems Using Event-Based Models of Behavior,” in *Proceedings of the ACM SIGPLAN/SIGOPS Workshop on Parallel and Distributed Debugging, ACM SIGPLAN Notices/*, 1989.
8. D. D. Clark, “The Structuring of Systems Using Upcalls,” in *#sosp10#*, 1985.
9. R. E. Griswold, C. L. Jeffery and G. M. Townsend, Graphics Programming in Icon, San Jose, California: Peer-to-Peer Communications, 1998.

Chapter 4
Visualizing Aspects of Program Behavior

This chapter demonstrates Unicon's execution monitoring facilities with examples of a variety of monitoring techniques. The examples are actual program fragments (rather than pseudocode) that show how to program various forms of monitoring. The examples all follow the common outline and monitoring facilities that were described in the preceding chapter. In particular, this chapter covers how to extract information in order to visualize:

- Control Flow
- Data Structures' Changes and Contents
- Memory Allocations, Collections, and Regions

4.1 Following the Locus of Execution

Perhaps the most basic monitoring act is following along in the source-code as execution progresses. Locus of execution information is used in various tools such as source-code viewers and profilers. Frequently, location information is used in combination with other execution information to inform the user of the specific source code line and column responsible for some behavior of interest.

This chapter presents simple example EMs that monitor location information and present it graphically. The first set of tools shows recent line number changes. These tools are primarily useful in detecting irregular control flow patterns that merit investigation, and in detecting major phases in program execution. Following the line number activity monitors, a graphical location profiler that displays cumulative location information is presented. Profilers are primarily useful in performance tuning.

The examples in this and the next several chapters are intended to demonstrate the broad capabilities of the monitoring framework. Actual source code is given in order to demonstrate useful techniques and affirm the claim that the framework supports an exploratory programming style. While the examples are often suggestive of monitors that are useful in their own right, they are necessarily kept simple

C. Jeffery, J. Al-Gharaibeh, *Writing Virtual Environments for Software Visualization*, DOI 10.1007/978-1-4614-1755-2_4

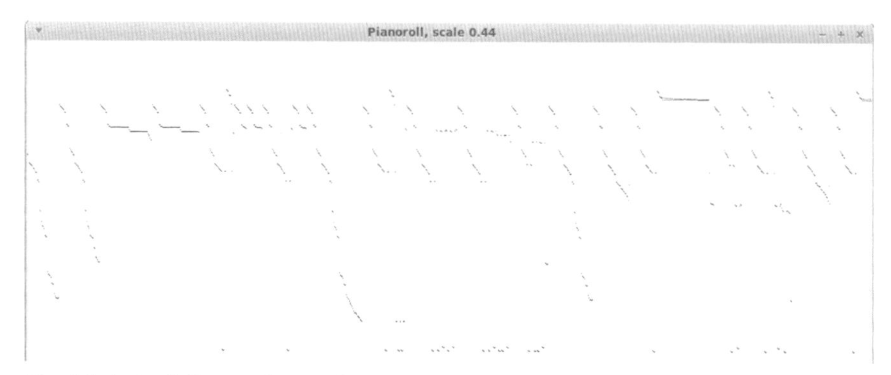

Fig. 4.1 A simple line number monitor

for exposition. The development of more sophisticated monitors is an open-ended research domain for future work that this framework was designed to facilitate.

4.1.1 Location Events

An event report with the code `E_Loc` occurs whenever the source line or column changes. Tracking the execution locus minimally involves selecting this event code in the event mask that is passed to `EvGet()` along with any others that may be of interest.

The value associated with a change in location is a 32-bit integer encoding of the line and column numbers. The line number is given in the least significant 16 bits, and the column number in the most significant 16 bits.

4.1.2 A Simple Line Number Monitor

The code segment that follows outlines a simple line number monitor that presents the sequence of source code lines on a strip chart. The y coordinate is used to denote the line number; successive line numbers are plotted adjacently along the x axis. Line numbers are scaled to fit the available screen space. A sample screen image is shown in Fig. 4.1. The tool is animated, showing the last n line number changes, where n is the width of the monitor window. As the animation progresses, ordinary sequential execution of successive expressions appears in the window as a downward sloping line. Periodic repetitions of patterns in the window indicate the execution of loops.

The EM starts by initializing the event monitoring system and opening a window on which to display its output. Local variables `x` and `y` refer to screen coordinates; `scale` is used to adjust the y coordinate to fit within the bounds of the window. Real numbers are used in the scaling arithmetic in order to use all of the available window space.

```
&window := open("LineMon", "g", "height=256", "width=256") |
  stop("can't open window")
scale := real(WHeight()) / prog_len()
x := 0
```

The program's main loop reads a location event with a call to `EvGet()`, computes and scales the line number to the window height, and plots it in the window with a call to `DrawPoint()`. After the point is plotted, `x` is advanced to plot the next line number in the next pixel column to the right. When the plot reaches the right edge of the window, the EM wraps around to the left edge. Because pixel columns are reused, a rectangle one pixel wide is erased at each iteration (`EraseArea()`'s height argument defaults to the entire window).

```
while EvGet(E_Loc) do {
  y := vertical(&eventvalue) * scale
  DrawPoint(x, y)
  x := (x + 1) % WWidth()   # advance x, wrap from right to left
  EraseArea(x, 0, 1)        # clear pixel column for next plot
  }
```

Variations on the line number monitor are presented in Figs. 4.2 and 4.3. Figure 4.2 draws a segment between the current source line and the preceding source line at each step. The effect emphasizes large jumps in program location that otherwise might not be noticed due to extremely short visits to certain locations. This phenomenon occurs more frequently in procedures that generate multiple results from a single expression than it does in ordinary procedural code. Figure 4.3 plots all the lines that execute in a single CPU clock tick (a hardware-dependent value; typically 4–20 ms) in a single column. This view compresses much more location information onto a single screen, but loses the ordering between specific location events within a clock tick.

4.1.3 A Location Profile Scatterplot

Another location monitoring example, presented below, renders a continuously updated scatterplot of program activity by source program line and column number. A sample screen image is presented in Fig. 4.4. The tool's animation does not employ motion, but rather changes in color as execution commences. The colors are rendered as grayscales for publication.

This EM maps source code columns and lines onto the x- and y- dimensions, one line or column per pixel. This mapping may be useful or already familiar to the user because it is a miniaturized view of the program text itself. Each source location at which the TP executes is highlighted, with the number of times that location has been executed given by a color progression on a logarithmic scale, from gray and blue

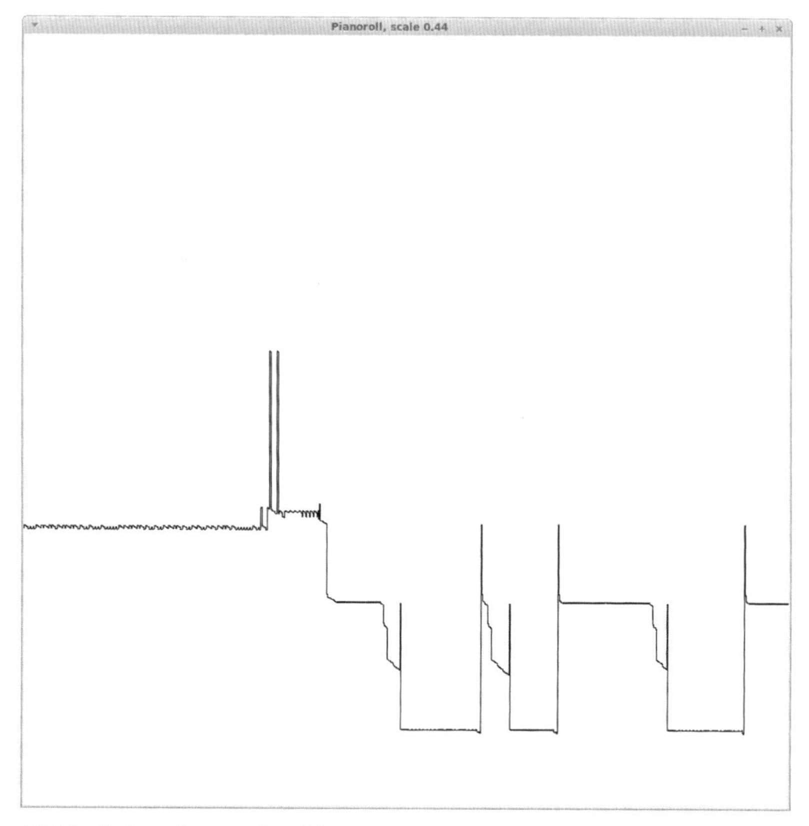

Fig. 4.2 Monitoring adjacent pairs of lines

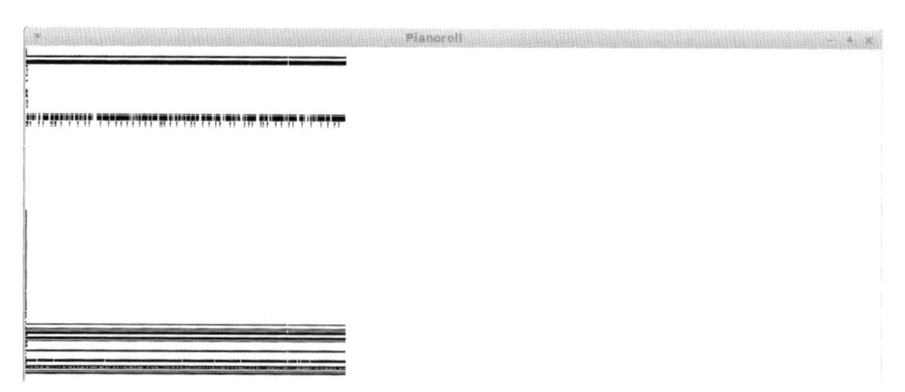

Fig. 4.3 Mapping CPU clock ticks to pixel columns

Fig. 4.4 A location profile scatterplot

through green and yellow and on to orange and red for locations that have executed many times.

The EM starts with standard initialization code and then creates a list of bindings with the various colors. A table, `counts`, maintains the number of times execution has occurred at each location.

```
&window := open("locus", "g", "bg=white", "size=80,500") |
   stop("can't open window")
Color := [ ]
every put(Color, Clone(&window, "fg=" || ("gray" | "blue" | "green" | "yellow" | "orange" | "red")))
counts := table(0)
```

With initialization completed, the main loop requests a location event, decodes its line and column, and increments the execution count for the location, stored in the table as `counts[&eventvalue]`. A point is then drawn in the window with a color encoding the log of the location's execution count. If the window height is not large enough to map the source file lines onto pixels, a bar is drawn at the bottom of the window to indicate that it has been clipped. A more sophisticated version of this program scales the mapping from lines to pixels.

```
while EvGet(E_Loc) do {
  y := vertical(&eventvalue)
  x := horizontal(&eventvalue)
  counts[&eventvalue] +:= 1
  value := integer(log(counts[&eventvalue], 6)) + 1
  if Context := Color[value] then
    FillRectangle(Context, x – 1, y – 1, 2, 2)
  If y > WHeight() then
    FillRectangle(0, WHeight() – 4, 80, 4)
  }
```

4.1.4 Tracking Source File Changes

Unicon programs can be separately compiled. The current source code file at any given point in execution is given by the keyword `&file`. In a monitoring situation, the current file in the monitored program can be obtained at any instant by querying the value of the `&file` keyword in the monitored program with the Unicon function `keyword()`:

```
source_file := keyword("file", Monitored_Program)
```

Checking a program's `&file` keyword is simple and easy, but as the frequency with which the monitor needs this information increases, it becomes equivalent to polling for state information changes. Since minimizing the computational overhead of monitoring can be important, some monitors find it worthwhile to use the event mechanism to reduce the overhead of maintaining source file information.

No explicit event informs programs of source file changes; it must be reconstructed from other events. In the absence of preprocessor directives to the contrary, the source file can only change when execution transfers into a different procedure. In that case, the current source file can be maintained using techniques described in the following chapter on monitoring procedure activity.

4.2 Monitoring Procedure Activity

Procedure activity is a major aspect of control flow, and it is especially significant in Unicon because procedures can generate more than one result. This section describes the monitoring of procedure activity in detail. The techniques presented are important because they also apply to the monitoring of Icon's built-in functions and operators as well as string scanning environments. The examples given are intended to illustrate the framework's capabilities and are by no means the best or only way in which procedure activity may be portrayed.

In order to model the semantics of generators, most EMs maintain trees of suspended procedure activations that may be resumed. After presenting techniques to maintain these trees, Sect. 4.2.2 describes an EM that draws an animated scatterplot of the number of results that each procedure produces; it quickly shows which procedures are generators, and shows when the number of results a procedure is producing changes significantly. Knowing which procedures are generators can be important for students and program maintainers that are unfamiliar with a program. For programmers that *are* familiar with the target program, knowing the number of results being produced per call to a given procedure can be valuable during debugging; it can confirm expected behavior and/or point out anomalies.

Section 4.2.3 presents an EM that gives an abstract view of the actual tree of active and suspended procedures; it is useful for understanding the path that control flow took to get to the current place of execution. This EM is generalized to include string scanning operations in Sect. 4.4.4.

As mentioned in the previous chapter, events take place at procedure calls, suspensions, resumptions, returns, failures, and implicit removals. The constant `ProcMask` contains a cset for all the event codes related to procedures; similar constants `FncMask`, `OperMask`, and `ScanMask` are used for other types of expression activity.

4.2.1 Activation Trees

The event value for calls and resumptions gives the procedure being activated, but other procedure events such as suspension and return give the value being produced. In order to track the currently active procedure, the monitor must maintain a model of the program's procedure *activation tree* (Fig. 4.5).

Fig. 4.5 An activation tree

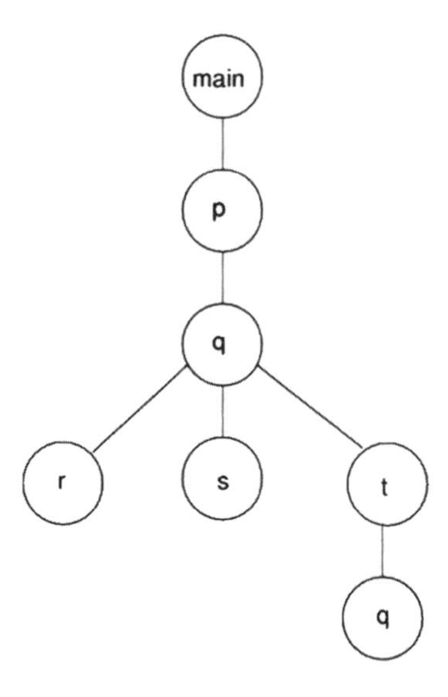

The procedure `evaltree()` described in this section maintains a simple model of procedure activation trees using records for tree nodes. Each record corresponds to an activation of a procedure. The record contains the procedure, the parent activation record from which the procedure was called, and a list of any children (including suspended ones) that this activation of the procedure has called:

```
record activation(value, parent, children)
```

When used in an EM, the record type may have additional fields to maintain other information about the procedure activation, such as the number of results it has produced. Fig. 4.6 shows the structures formed by `evaltree()` to model the activation tree in Fig. 4.5. The source code for `evaltree()` may be found in `ipl/mprocs/evaltree.icn` in the Unicon language distribution.

The procedure `evaltree()` maintains the complete activation tree as well as the current activation with the following monitor event loop. It is called with an event mask parameter and two procedure parameters. The event mask parameter gives all the events needed by the EM. The procedure parameters consist of a callback procedure used to inform the monitor of changes in the tree, and a record constructor for a record type that has at least the fields declared above. The callback procedure is called with the activation record being entered as well as the activation record being exited.

```
procedure evaltree(mask, callback, activation_record)
  # compute codes for each branch of the case clause from mask
  while EvGet(mask) do
    case &eventcode of {
      # maintain activation tree, call client callback procedures
      }
end
```

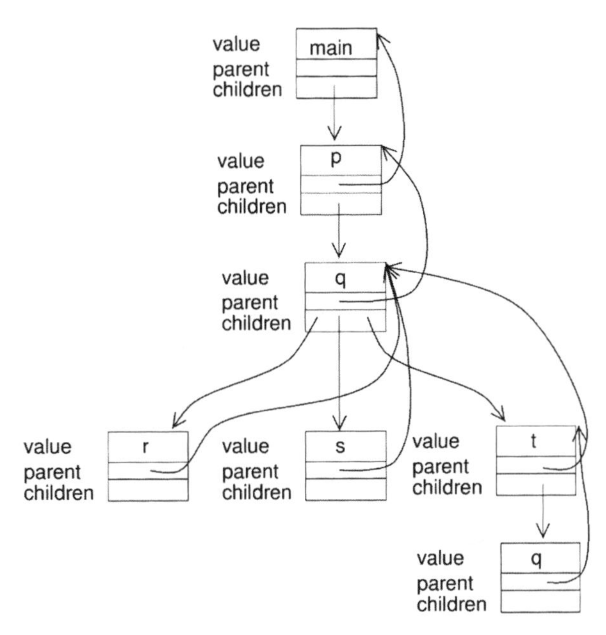

Fig. 4.6 A Unicon representation of an activation tree

In order to operate properly with any combination of procedure, function, operator, and scanning environment events, `evaltree()` examines its event mask and builds up lists of codes related to each of the six tree-modifying events. It stores these lists in the global variables `CallCodes`, `SuspendCodes`, `ResumeCodes`, `ReturnCodes`, `FailCodes`, and `RemoveCodes`. In addition, `evaltree()` creates a dummy root activation on which to build the activation tree.

The branches of `evaltree()`'s case clause perform the actual tree manipulations and then call the client callback procedure, supplying it with both the activation being entered and the activation being exited. For each call event, a new node is created and inserted as the rightmost child of the current node. The new node becomes the currently executing node.

```
!CallCodes: {
entered := activation_record()
entered.node := &eventvalue
entered.parent := current
entered.children := [ ]
put(current.children, entered)
current := entered
callback(current, current.parent)
}
```

Return and fail events result in the inverse of a call event: The current node is removed from the activation tree, and the parent of the current node becomes active. When an Unicon `return` expression is executed, the instrumentation produces removal events for all descendants of the returning node preceding the resulting return event.

```
!ReturnCodes | !FailCodes: {
    exited := pull(current.parent.children)
    current := current.parent
    callback(current, exited)
    }
```

Suspend and resume events do not change the structure of the tree. For suspend events, the parent becomes the current (active) node; for resume events, the rightmost suspended child is resumed and becomes the current node. After the current node is updated, the client callback procedure is called.

```
!SuspendCodes: {
    current := current.parent
    callback(current, current.children[ –1])
    }
!ResumeCodes: {
    current := current.children[ –1]
    callback(current, current.parent)
    }
```

Removal events denote the implicit exit of a node in the activation tree as a result of control flow. Typically, a removal event precedes the current node's return or failure and denotes the destruction of the current node's rightmost child. If the current node has no children, removal indicates an implicit destruction of the current node, indicating that it will not be used in the surrounding expression evaluation context.

```
!RemoveCodes: {
      if exited := pull(current.children) then {
        while put(current.children, pop(child.children))
        callback(current, exited)
        }
      else {
        exited := pull(current.parent,children)
        current := current.parent
        callback(current, exited)
        }
      }
```

The `default` clause in this case expression simply calls the client callback procedure. The activation tree is not modified. This clause is useful because execution monitors that use `evaltree()` may be interested in other types of events besides those that involve the activation tree.

```
default: callback(current, current)
```

4.2.2 An Animated Call-Result Scatterplot

The following example illustrates the use of `evaltree()`, and introduces a simple library for programs that draw animated scatterplots, `scatlib`. This example plots the number of times each procedure has been called along the x axis, while the number of results it has produced is plotted along the y axis. Points are moved whenever either a call or a resumption occurs. Red is used for user-defined procedures, while green indicates activity for less expensive built-in functions. If the user presses a mouse button on one of the plotted points, the names of any procedures plotted at that point are listed. An example screen image from this program is given in Fig. 4.7; the name `GenMoves` in the lower-right corner is the name of the procedure plotted at the last location on which the mouse was clicked. The image does not convey the nature of the animation, in which plotted points start in the upper-left corner and migrate down and to the right at varying speeds and directions.

A call-result scatterplot serves several purposes. It serves as a basic procedure call profiler, revealing which procedures are used the most and are therefore most important in overall performance. Since this information is presented while the program is executing, it provides quicker feedback than profilers that present information only after execution has run to completion. Feedback during execution also shows temporal changes associated with major phases in the program. These uses are language-independent. The call-result scatterplot also serves two language-specific purposes: it shows the user which procedures are generators, and how many results the procedures are producing per call.

When a procedure consistently produces no results, it moves horizontally along the top edge. On the other hand, if a procedure generates results, it moves vertically straight down. If a procedure consistently returns with one result, it moves diagonally down and across. The slope of a line from the origin to a given procedure's point on this graph gives the average number of results that procedure has produced per call. If the motion of a point plotted for a procedure changes its direction substantially, it may indicate unusual behavior that is worth further examination.

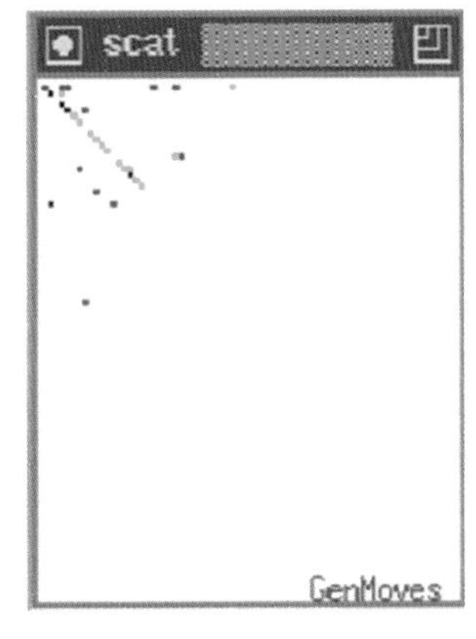

Fig. 4.7 A scatterplot with motion

4.2.2.1 Scatlib

The `scatlib` library provides animated scatterplot capabilities through a simple interface. The primary function of the library is to track a moving collection of objects that are mapped onto x, y coordinates using a user-supplied scale. Multiple objects may be plotted to the same coordinate. Objects may be plotted using different colors.

4.2.2.2 The Call-result Scatterplot Application

Two global tables, `xcounts` and `ycounts`, store the call and result counts for each TP procedure. The global table `at` maintains a set of objects plotted at each point on the graph; `at` is keyed by the integer-encoded locations introduced in the preceding chapter and is discussed in more detail later.

```
global  at,        # table of sets of objects at various locations
        xcounts,   # table of x counts
        ycounts,   # table of y counts
```

Procedure `main()` initializes the execution monitoring framework, opens a window, and initializes the scatterplot library by calling `scat_init()`, and hands off the program's flow of control to `evaltree()`. `evaltree()` in turn obtains events, builds the activation tree, and calls `scat_callback()` for each event report. `main()` passes `scat_callback()` to `evaltree()` as a parameter, in addition to the event mask to use and the record type to use for activations. The event mask includes procedure events selected by the symbol `ProcMask` and monitor interaction events, indicated by the symbol `E_MXevent`. Monitor interaction events, described in Chap. 3, provide a convenient means of incorporating user input such as mouse clicks and button presses into EMs without a need for separately polling the EM window for activity.

```
# ... from procedure main()
&window := open("scat", "g", "width=150", "height=180") |
  stop("can't open window")
xcounts := table(0)
ycounts := table(0)
at := table()
evaltree(ProcMask ++ E_MXevent, scat_callback, activation)
```

Procedure `scat_callback()` updates the plotted location of a procedure whenever it is called or produces a result, calling `plot()` to increase the appropriate procedure's x- or y-coordinate, respectively. If the event is a call, the point corresponding to parameter `new` (the activation being entered) is updated, while if the event is a suspend or a return, the point corresponding to parameter `old` (the activation being exited) is updated.

If the event indicates user activity, the user input is supplied in `&eventvalue`, and the keywords `&x` and `&y` are updated with the mouse location. If the user presses the escape character `"\e"`, monitoring is terminated; if the user presses a mouse button, `write_names()` is called to write the names of procedures plotted where the mouse indicates.

```
procedure scat_callback(new, old)
  case &eventcode of {
    E_Pcall:                plot(new.node, 1, 0)
    E_Psusp | E_Pret:         plot(old.node, 0, 1)
    E_MXevent: {
      case &eventvalue of {
        "\e": stop("execution halted")
        &lpress: repeat {
          write_names()
          if Event() === &lrelease then break
          }
        }
      }
  }
end
```

The procedure `plot()` takes a procedure and updates the tables to reflect its new position. If the procedure is the only occupant of the screen coordinate it is leaving, the point is erased there; similarly, if the new position is not already occupied, a point is drawn. *Points* are plotted two pixels wide and two pixels high because individual pixels provide poor visibility on some displays. An even larger size might improve visibility further at a cost of screen space. `plot()` uses a logarithmic scale in order to keep the screen size required by this application reasonable for large programs. A logarithmic scale is chosen over a linear scale because any linear scale would either plot the most important often-called procedures off the edge of the chart or else plot all the less frequently called functions together in one corner of the chart. The scaling process uses the distance of the point from the origin in order to preserve the ratio of calls to results in the scaled point; this is discussed in more detail below.

```
procedure plot(who, iscall, isrslt)
  loc := scaled_location(xcounts[who], ycounts[who])
  if *delete(\at[loc], who) = 0 then
    EraseArea(horizontal(loc) * 2, vertical(loc) * 2, 2, 2)
  xcounts[who] +:= iscall
  ycounts[who] +:= isrslt
  loc := scaled_location(xcounts[who], ycounts[who])
  /at[loc] := set()
  if *insert(at[loc], who) = 1 then
    FillRectangle(horizontal(loc) * 2, vertical(loc) * 2, 2, 2)
end
```

Procedure `scaled_location(x, y)` scales its arguments and produces an integer encoding of the point (x, y) with the x-coordinate in the most significant 16 bits and the y-coordinate in the least significant 16 bits. `scaled_location()` also computes the distance from the origin for a point using the Pythagorean theorem; it is used during scaling.

```
procedure scaled_location(x, y)
   length := sqrt(x ^ 2 + y ^ 2)
   return location(scale(y, length), scale(x, length))
end
```

The procedure `scale(coord, len)` applies a logarithmic scaling factor to a coordinate. If logarithmic scales were applied separately to the x- and y-coordinates, the proportions of calls to results would not be preserved and the resulting points would be plotted artificially close to the central diagonal of slope 1. Instead, the logarithmic scale is applied to the distance from the origin. The coordinate is multiplied by the ratio of the scaled length to the original length. When both coordinates are so scaled, the scaled point forms a similar triangle to the original unscaled point; the slope of calls to results is preserved from the unscaled point.

```
procedure scale(coord, length)
   if length <  1 then return 0          # avoid divide by 0 error
   return integer(coord * log(length, 1.25) / length)
end
```

Procedure `write_names()` prints the names of all procedures plotted near a mouse click. It builds a list `L` of the names of all procedures in the `at` table located within one pixel of the current mouse location. When `write_names()` has built the list of procedures, it erases the last name list and writes the new list of names in the lower-left corner of the window.

```
procedure write_names()
static maxrows, maxcolumns
  &x /:= 2
  &y /:= 2
  # build a list of names of procedures
  L := [ ]
  every i := - 1 to 1 do
    every j := - 1 to 1 do {
      loc := location(&y + j, &x + i)
      every put(L, procedure_name(! \ at[loc]))
      }
    # compute the geometry needed to erase last name list
    if max := *L[1] then {
      every max <:=*!L
      maxcolumns <:= max
      }
    maxrows <:=*L
    &col := WColumns() - maxcolumns
    &row := WRows() - maxrows - 1
    EraseArea(&x, &y)
    if *L > 0 then
      every i := 1 to *L do {
        GotoRC(WRows() - *L + i, WColumns() - max)
        writes(&window, L[i])
        }
    e := Event()
end
```

The `scat` program could be generalized in several ways; for instance, it is trivial to extend `scat` to accommodate Unicon's built-in function and operator repertoire. If this information were cross-referenced with static knowledge of which functions and operators are generators, `scat` could show whether they are being used generatively, or only used to obtain single results as in conventional programming. Another useful way to extend `scat` would be to allow the user to specify lines (slopes) to indicate a procedure's expected result/call ratio; if the number of results were too low or too high, the user might want to stop execution and inspect the situation in closer detail.

4.2.3 Algae

A program named Algae illustrates one approach to displaying procedure and generation activity in a more connected fashion. Algae displays an animated representation of the activation tree for procedures, built-in functions, and/or string scanning environments as the TP executes, and serves as a basis for other more sophisticated EMs that are presented in later chapters.

Algae is designed to use little screen space and does not require rearrangement of nodes as the tree changes, unlike conventional approaches to tree layout. This attempt to save screen space and animation time produces an approximation of the activation tree that sacrifices the details of parent-child relationships in the tree. The Algae metaphor is meant to complement more conventional layouts, not to replace them. The idea behind Algae is to present enough of the expression activity so that common goal-directed evaluation patterns in TP are identified and strange behavior can be noticed as an unfamiliar pattern in the animation.

4.2.3.1 Algae Geometry

The Algae window uses a simple two-dimensional grid of cells; the vertical dimension depicts expression nesting depth, such as calls and returns from procedures. The horizontal dimension depicts generator suspension width, such as procedure, function/operator, and scanning environment suspension. Whenever a computation is suspended, new computations at the same level start in the next cell column to the right, indicating the possibility of backtracking into the suspended computation. A sample image of Algae is shown in Fig. 4.8. The target program being monitored is a recursive descent parser. Magenta (or dark gray) cells represent suspended procedures for the nonterminals of a parse that is being attempted. A yellow (light gray) cell in the bottom right is the currently active procedure. Light blue (medium gray) is used to fill in cells when they are vacated; coloring these cells provides a high water mark for the computation up to any given point and gives it an overall characteristic shape.

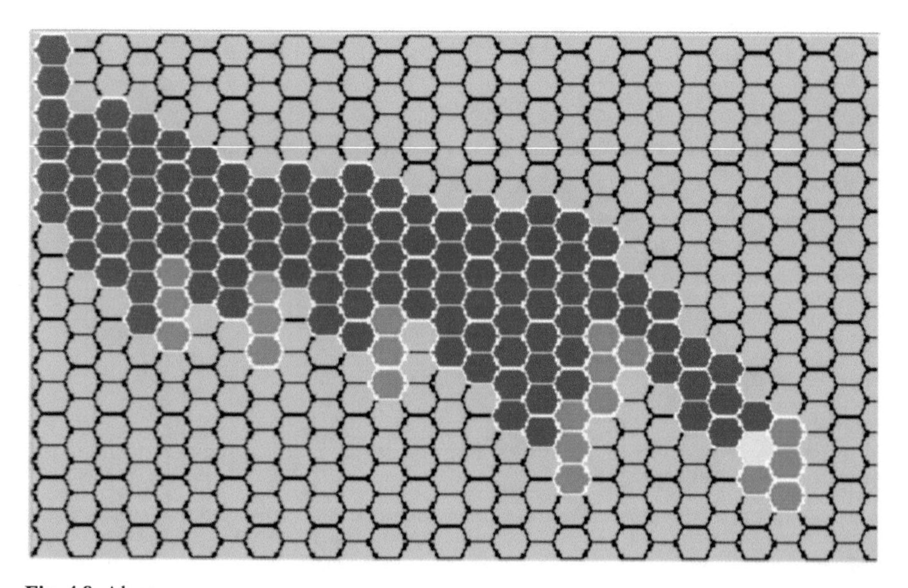

Fig. 4.8 Algae

In order to support the two-dimensional geometry, Algae's activation tree records have fields for the row and the column of the cell assigned for each activation:

```
record algae_activation(value, parent, children, row, column, color)
```

Since screen space is limited, each activation is depicted as a small hexagon in the window, color-coded by the kind of activation (procedure, function, operator, or string scanning environment). The size of the hexagons is scalable. Given this geometry, it would be easier to plot Algae using rectangular points. Hexagons are used primarily for their visual effect—they provide a smoother animation as the tree grows and shrinks. Position changes in Algae are often diagonal, and in a square mapping, these changes appear to be a farther distance than horizontal or vertical position changes.

`hexlib.icn` is a collection of procedures totalling roughly 160 lines that supports the manipulation of hexagon maps; it is omitted here for the sake of brevity. In the code below, the hexlib procedure `spot(window, row, column)` fills a hexagon at a given location with a particular color.

Because screen space is limited and the activation tree is constantly changing, Algae does not lay out the tree in a way that spreads out nodes throughout the available screen space. Instead, Algae lays out tree nodes from the leftmost edge of the window, being careful to maintain the correct depth and breadth of the tree, and making sure that no two nodes occupy the same cell. When a new node is created, it is a assigned a cell with a row given by its level; the column is computed by inspecting the existing tree and finding the first position to the right of both the parent node and any nodes at the new node's level.

Since expression trees grow and shrink along their rightmost edge, the tree search to assign a column is a preorder, depth-first, right-to-left search. An important special case is if the node's parent already has a child, in which case the newly created node can immediately be assigned a column adjacent to its older sibling; this case is handled directly in `algae_callback()` for efficiency, and often allows the tree search to be avoided entirely.

The code to compute the column is:

```
procedure computeCol(parent)
  node := parent
  while node.row > 1 do node := node.parent # find root
  if node === parent then return parent.column
  if col := subcompute(node, parent.row + 1) then
    return max(col, parent.column)
  else
    return parent.column
end

procedure subcompute(node, row)
  # check this level for correct depth
  if \ node.row = row then return node.column + 1
  # search children from right to left
  return subcompute(node.children[*node.children to 1 by – 1], row)
end
```

4.2.3.2 Using evaltree() to Incrementally Update the Display

Algae makes extensive use of colors to indicate the kind of activation, such as whether it is a procedure, function, or string scanning environment. In `main()`, several bindings are created with different foreground colors, as described in Chap. 3. The colors used are arbitrary and the user can determine the contents of the node by clicking on it if the color is not familiar.

After initialization, Algae calls `evaltree()` and passes it a reference to the procedure `algae_callback()`. The event mask used is variable and depends on command-line arguments. The body of `algae_callback()` performs the incremental animation of the tree. Each event that modifies the activation tree entails the updating of two display cells: a cell that is entered is drawn in yellow to mark it as the active cell, and a cell that is exited is either drawn in the color associated with the activation (if it is suspended) or in a background gray color (if the associated activation has returned or failed and no longer exists).

```
case &eventcode of {
  !CallCodes: {
    new.column := (old.children[ - 2].column + 1  | computeCol(old))
    new.row := old.row + 1
    new.color := Color[&eventcode]
    spot(\old.color, old.row, old.column)
    }
  !RetCodes | !FailCodes: spot(background, old.row, old.column)
  !SuspCodes | !ResumCodes: spot(old.color, old.row, old.column)
  !RemCodes: {
    spot(black, old.row, old.column)
    WFlush(black)
    delay(100)
    spot(background, old.row, old.column)
    }
  E_MXevent: user_event(&eventvalue, new)
  }
spot(yellow, new.row, new.column)
```

4.2.3.3 Algae Controls

User control of Algae consists of marking specific hexagons (using the left mouse button) or entire rows and columns (using the middle button) to pause execution. Pressing the right button atop a hexagon marked active or suspended prints the name of the associated procedure or function, or the subject of the associated string scanning environment. The input handling is performed by `do_event()` in response to an `E_MXevent`.

Each call to `algae_callback()` checks to see whether the cell being entered is one selected by the user to pause execution, and if it is, the callback procedure loops, reading user events until the user indicates that execution should continue. `algae_callback()` concludes with the code for this test:

```
loc := location(new.row, new.column)
if \ step | (new.column >= \ maxcolumn) | (new.row > = \ maxrow) |
  \ hotspots[loc] then {
  step := &null
  WAttrib("windowlabel=Algae stopped: (s)tep (c)ont ( )clear ")
  while e := Event() do
    if user_event(e, new) then break
  WAttrib("windowlabel=Algae")
  }
```

The procedure `user_event()` returns if execution should proceed, but fails if execution is still paused and another user event should be obtained. The code for `user_event()` is somewhat lengthy and is included in the complete text of Algae in `ipl/mprogs/algae.icn` in the Unicon language distribution.

4.2.4 Maintaining the Current Source File

The preceding chapter showed a polling technique for obtaining the current source file name in which execution is taking place. Polling is expensive, and for ordinary programs it is better to maintain source file information by tracking the current procedure. The computational load imposed by monitoring can be minimized by only querying `&file` the first time each procedure is called and maintaining these results in a table. The combined code looks like:

```
source_files := table()
while EvGet(ProcMask) do {
  case &eventcode of {
    E_Pcall: {
      / source_files[&eventvalue] := keyword("file", Monitored)
      current := activation(&eventvalue, current, [ ])
      put(current.parent.children, current)
      current_file := source_files[current.p]
      }
    E_Psusp: {
      current := current.parent
      current_file := source_files[current.p]
      }
    E_Presum: {
      current := current.children[ – 1]
      current_file := source_files[current.p]
      }
    E_Pret    E_Pfail: {
      pull(current.parent.children)
      current := current.parent
      current_file := source_files[current.p]
      }
    E_Prem: {
      child := pull(current.children)
      current.children ||| := child.children
      }
  }
}
```

The techniques presented in this section apply not only to procedures, but also Unicon's built-in functions, operators, and scanning environments—the `evaltree()` procedure can accommodate all of these kinds of events simultaneously and maintain one large expression activation tree. Some differences between the different kinds of activations exist; an obvious one is that function and operator events are so frequent that monitoring them in an EM like Algae vastly reduces the tool's effectiveness in monitoring the less frequent procedure activity. It would be useful to explore variants of `evaltree()` that allow certain subtrees to be ignored, or to not plot activity at all unless interesting behavior such as generation or backtracking takes place.

4.3 Monitoring Memory Usage

In Unicon, memory management is automatic. Memory is allocated when it is needed, and reclaimed implicitly when it is no longer referenced. The memory management subsystem provides significant insight into program behavior. Program performance problems can often be attributed to inefficient memory usage, and the actual pattern of usage can illuminate aspects of behavior ranging from simple transitions between major phases of the program down to semantic errors in program coding.

Memory usage is interesting to study because it is not directly evident from source code examination. The execution monitoring instrumentation produces events on every memory allocation with an event code that indicates the type allocated and a corresponding event value giving the size of the allocation in bytes. In addition, events occur at garbage collections, including the types and sizes of objects that survive reclamation. Allocation events are selected with the `evinit` symbol `AllocMask`.

This section presents a variety of EMs that portray aspects of memory usage. First, EMs are given that plot each individual allocation in relation to other recent allocations; they are useful in observing localized program behavior such as allocations of unusual size or changes in the major phases of execution. Later in the chapter, EMs that portray cumulative memory usage behavior are discussed; they provide a useful profiling service and a general understanding of the TP's use of memory. These simple examples illustrate only a few of many visual metaphors that have been developed for memory usage, ranging from literal views of the heap to completely abstract animations whose patterns reflect a program's memory allocations. Some of the other tools that portray memory activity are described in a separate document [1].

4.3.1 Allocation by Type

Many visual metaphors can be used to depict allocation types or sizes, or both. Two allocation monitors are presented in this section. The first emphasizes frequencies and patterns of types in allocated memory, while the second emphasizes allocation size information. These examples also exhibit a clean separation of the data collection and graphics rendering tasks, enabling the visual metaphors to be used in other tools that monitor types of events other than memory allocations.

4.3.1.1 Pinwheel

The pinwheel metaphor presents a sequence of values, in this case the event codes associated with allocation event reports, encoded as colors or textures drawn in sectors around a circle. The *n* sectors of the circle represent a history of the last *n* allocation events in the TP's execution. A screen image from a program using this

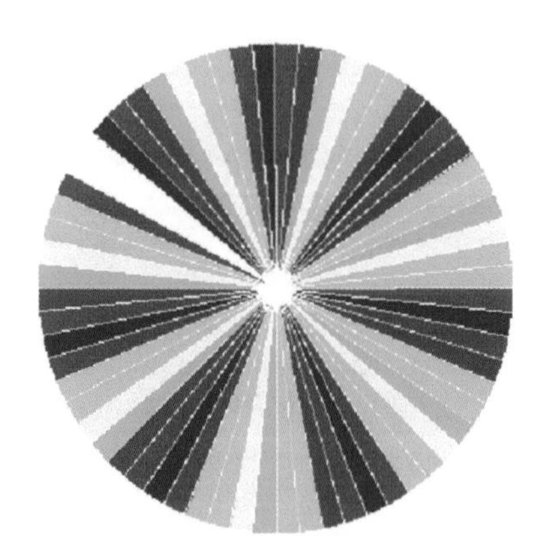

Fig. 4.9 Pinwheel

metaphor to present memory allocation patterns is given in Fig. 4.9. In this example, event codes for Unicon's allocated types are mapped onto colors. The view is updated on each allocation; the animation rate gives an indication of the frequency with which memory allocations occur.

Pinwheel and many other visual metaphors have been encapsulated in procedures for use by execution monitors. By using a common set of conventions, the metaphors can be applied interchangeably and to different types of data. The procedure `pinwheel()`, called with no arguments, starts with local variable declarations and then initializes several variables that scale the mapping.

```
procedure pinwheel()
local clear, xorg, yorg, radius, radians
local angle, arc, sector_units, fullcircle, blank, max, xratio, yratio
  max := real((WWidth() < WHeight()) | WWidth())
  xratio := WWidth() / max
  yratio := WHeight() / max
  fullcircle := 2 * &pi
  radians := 0
  sector_units := fullcircle / Sectors
  blank := 2 * sector_units
  xorg := WWidth() / 2
  yorg := WHeight() / 2
  radius := max / 2
  while NextEvent() do {
    FillArc(Background, 0, 0, WWidth(), WHeight(), radians + sector_units, blank)
    FillArc(Binding, 0, 0, WWidth(), WHeight(), radians, sector_units)
    DrawLine(Background, xorg, yorg, xratio * radius * cos(radians)
             + xorg, yratio * radius * sin(radians) + yorg)
    radians +:= advance
    }
end
```

Pinwheel's main loop reads a monitoring event, draws a filled arc in a binding that uses a color associated with the event, and erases the next slice of the pinwheel to mark the edge of motion. The local variable `angle`, the front edge of the pinwheel motion, is advanced at each iteration. The procedure `NextEvent()` encapsulates the task of reading a program event and selecting an appropriate color (or texture) to portray it so that the type of data being processed and the color used to draw the pinwheel are independent of the task of drawing the pinwheel itself. `NextEvent()` assigns the global variable `Binding` a window value with an appropriate foreground color for use in drawing the sector.

4.3.1.2 Nova

The nova metaphor is another example of a radial mapping of a sequence of event reports. Each allocation event report is plotted as a line segment from the center of the window in polar coordinates, with a radius given by the size of the allocation (`&eventvalue`), at a regular angular offset from the preceding value. Like pinwheel, the graphic is drawn in a color that indicates the allocation type, based on the event code, and the display is animated at the rate at which memory allocations take place. An example screen image from nova is shown in Fig. 4.10.

Like pinwheel, nova begins with an initialization section, followed by a loop that reads an event (again using `NextEvent()`) and draws a line at the appropriate angle and of the appropriate length.

```
procedure nova()
local clear, xorg, yorg, radius, radians
local arc, sector_units, fullcircle, erase, oldvalue
initial gclear := 1
  erase := list(Sectors)
  fullcircle := 2 * &pi
  radians := 0
  sector_units := fullcircle / Sectors
  xorg := WWidth() / 2
  yorg := WHeight() / 2
  radius := ((WHeight() < WWidth())| WHeight()) / 2.0
  while NextEvent() do {
    put(erase, Value)
    oldvalue := get(erase)
    DrawLine(Background, xorg, yorg, \ oldvalue * cos(radians) + xorg,
                oldvalue * sin(radians) + yorg)
    DrawLine(Binding, xorg, yorg, Value * cos(radians) + xorg,
                Value * sin(radians) + yorg)
    radians +:= advance
    }
    end
```

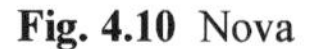

Fig. 4.10 Nova

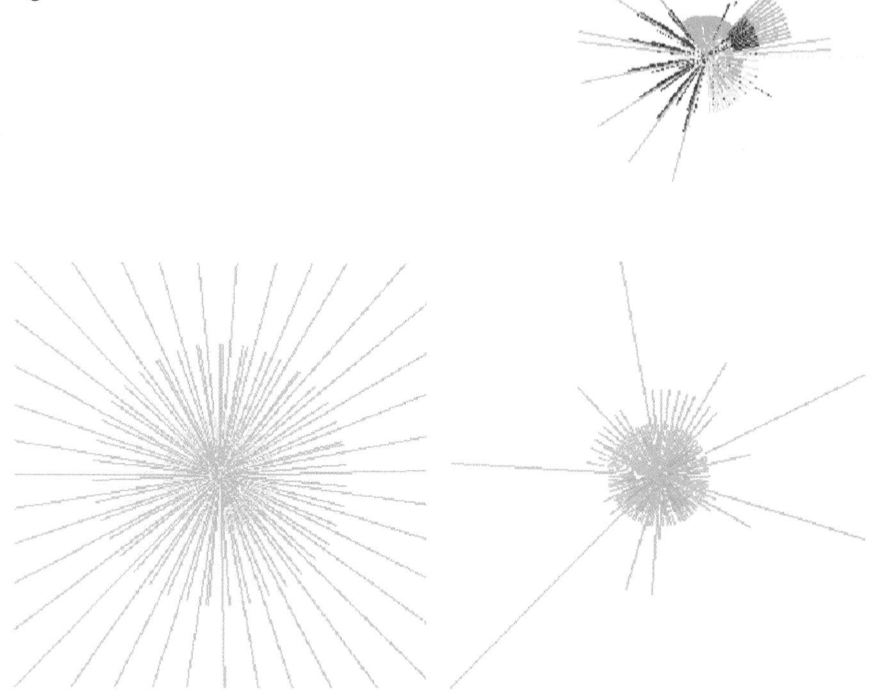

Fig. 4.11 Frequent large allocations suggest a problem. The program runs twice as fast after a two-line change

The following example demonstrates how memory allocation monitors may be of practical use. A poetry-scrambling program submitted by a user produced the visual signature given in Fig. 4.11 (left) when run under a tool using the nova metaphor. The wedge-shaped gap in Fig. 4.11 (left) is present simply because the nova's sweep has not completed its first revolution. The program builds up very long lists by repeated concatenation, resulting in the frequent very large allocations shown in the figure. After changing two lines of code to replace a list concatenation with calls to Unicon's `put()` function, the visual signature became normal, and program execution speed doubled (Fig. 4.11, right).

4.3.2 *Cumulative Allocation by Type*

Visualizing individual allocation events is useful for understanding local phenomena, but an overall summary of memory allocation is also useful in understanding program behavior. The following code segment totals the amount of memory allocated in the program by data type, building a table of sums that is keyed by the allocation event codes for each type. The sums are cumulative, that is, garbage collections are not taken into consideration.

```
t := table(0)
while EvGet(AllocMask) do
   t[&eventcode] +:= &eventvalue
```

4.3.2.1 Animating a Bar Graph

The following procedure renders a list of nonnegative numbers in a window as a bar graph. Each bar in the graph is given a string name in a list called `labels` and is drawn using a color from a list of color contexts named `colors`; the indices of `labels` and `colors` match those of the list of numbers. The example can be improved in many ways; the scale can be labeled more clearly, the origins may be supplied as parameters instead of computed from the data, and so forth.

```
procedure bar_graph(L, labels, colors, scale)
local height, x, y, i
  EraseArea()
  height := WHeight()
  bar_width := real(WWidth()) / *L
  WAttrib("label=Bar Graph, scale " || left(scale, 6))
  every i := 1 to *L do {
    x := (i − 1) * bar_width
    y := L[i] * scale
    FillRectangle(colors[i], x, height − y + 1, bar_width − 2, y)
    DrawString(x, 15, labels[i])
    }
end
```

If `bar_graph()` is called frequently, such as every time an event occurs in an execution monitoring setting, the frequent window updates create a distracting amount of screen flicker. In such an animation, an incremental approach is more appropriate.

The following program updates a bar graph incrementally. The bar graph presents cumulative memory allocation by type. An example screen image from this animated bar chart is given in Fig. 4.12.

The cumulative allocations are stored in list `bars`, in the order they appear on the screen. A parallel list of labels for each bar is maintained in `labels`; it is built from a table `evs` that maps event codes to their string names. The table is constructed by the standard `evinit` library procedure `evsyms()`. The mapping from event codes to screen position is maintained by the table `typecode2bar`. The animated bar graph scales itself as cumulative allocations increase.

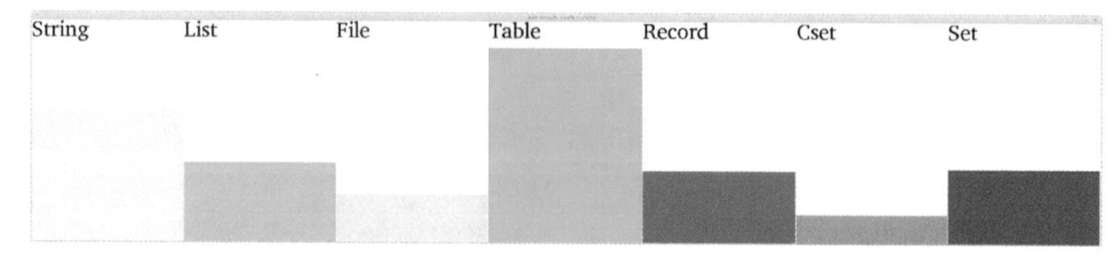

Fig. 4.12 An animated bar graph

```
&window := open("barmem", "g") | stop("can't open window")
height := WHeight()
evs := evsyms()
typecode2bar := table()
bars := [ ]
labels := [ ]
scale := 4.0
```

The main loop requests an allocation event and calls procedure `bar()` to update the size of the bar that corresponds to the event. A new bar is created when a type's first allocation takes place. No screen space is devoted to types for which no allocation occurs. As each bar's label is obtained from the event names table `evs`, the event's `E_` prefix is stripped by the string subscript `[3:0]`.

```
while EvGet(AllocMask) do {
  if /event2bar[&eventcode] := *put(bars,0) then {
    put(labels, evs[&eventcode][3:0] | "?")
    put(Colors, contexts[&eventcode])
    }
  extent := bars[event2bar[&eventcode]] +:= &eventvalue
  extent *:= scale
  if extent > height - 20 then bar_graph(bars, labels, Colors, scale /:= 2)
  else
    bar(extent, Colors[t[&eventcode]], event2bar[&eventcode])
  }
```

The procedure `bar()` simply fills in a rectangle for the added space.

```
procedure bar(extent, Color, i)
  x := (i - 1) * bar_width
  y := height - extent + 1
  FillRectangle(Color, x, y, bar_width - 2 , &eventvalue * scale + 1)
end
```

4.3.2.2 Pie Charts

The following procedure draws a pie chart from a table `shares` in which each portion of the pie represents a key and their relative size is the key's table value. A parallel table `colors` of window bindings contains the color, grayscale, or texture that is used to distinguish each of the parts.

```
procedure draw_pie(shares, colors, sum, x, y,width, height)
local start := 0, fraction := 360 * 64.0 / sum, k, path
  every k := key(shares) do {
    path := fraction * shares[k]
    FillArc(colors[k], x, y, width, height, start, path)
    start +:= path
    }
end
```

Unless the update rate is high, a visualization tool using this procedure can be animated by brute force by redrawing the entire image each time rather than incrementally. If the update rate is high, the chart might only be redrawn when a constituent's size changes by a significant amount, such as more than one percent of the total. A sample screen image from such a program is given in Fig. 4.13.

4.3.3 Running Allocation by Type

In order to take garbage collections into account, the program must select `E_Collect` and `E_EndCollect` events. The `E_Collect` event is produced prior to a garbage collection. The `E_EndCollect` event occurs after a garbage collection, and if it is selected, the monitoring instrumentation also produces (re)allocation

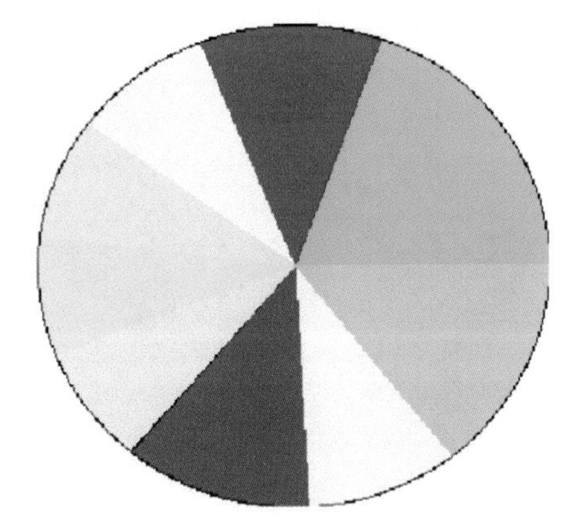

Fig. 4.13 A pie chart

events in between the `E_Collect` and `E_EndCollect` for the objects that survived the collection.

```
codes := AllocMask ++ E_Collect ++ E_EndCollect
t := table(0)
while EvGet(codes) do
  if &eventcode === E_Collect then t := table(0)
  else t[&eventcode] +:= &eventvalue
```

A more complex example of monitoring allocation by type is the following strip chart. It uses the same approach as the preceding example, but portrays a continuous animation in a window. In the following example, the y axis is used to show the proportions of memory used by all types. An example screen image from this program is given in Fig. 4.14.

The program monitors all memory allocation and garbage collection information, maintains a table of running sums of memory by type, and draws each vertical line in the graph as a set of segments that are color-coded by type and whose length corresponds to the proportion of memory used by that type. A library procedure named `typebind()` is linked and used to provide the color encoding. `typebind()` returns a table whose keys are type allocation event codes and whose values are window bindings with foregrounds set to various colors; the table is stored in global variable `Colors`. Since colors vary from device to device, several palettes are available from `typebind()`, depending on the output device to be used. The global variable `tallies` refers to a table of sums of allocations keyed by type. Global variable `heapsize` stores the total amount of available memory. The event processing loop in procedure `main()` calls `redraw()` to update the window on each allocation and clears the window on garbage collection.

Fig. 4.14 A memory allocation strip chart

```
tallies := table(0.0)
heapsize := 0
every heapsize +:= keyword("regions", Monitored)
&window := open("MemoryType", "g") | stop("can't open window")
Colors := typebind(&window, AllocMask)
mask := AllocMask ++ E_Collect
while EvGet(mask) do
  case &eventcode of {
    E_Collect: {
      EraseArea()
      tallies := table(0.0)
      }
    default: {
      tallies[&eventcode] +:= &eventvalue
      redraw()
      }
    }
```

The procedure `redraw()` updates the display when needed. Real arithmetic is used to minimize numeric errors in the mapping.

```
procedure redraw()
static x
initial x := 0
  start := 0
  every k := key(t) do {
    segment := WHeight() * real(tallies[k]) / heapsize)
    FillRectangle(Colors[k], x, start, 1, segment)
    start +:= segment
    }
  x := x + 1 % WWidth()
  EraseArea(x + 1, 0, 1)
end
```

It is possible to substantially improve on this trivial example; redundant calls and type conversions can be avoided, and many variations on the mapping from the problem space onto the image geometry are possible. In particular, it may be worth avoiding screen updates when the change to be reported is very small.

4.3.4 Survival Rates Across Collections

If a garbage collection reclaims only a small amount of storage, the TP may quickly run out of free memory and collect again. As the frequency of collections rises, overall system performance declines rapidly. This information can be obtained by selecting `E_Collect` and `E_EndCollect` events and reading TP's `&storage` keyword.

```
while EvGet(E_Collect) do {
  L := [ ]
  every put(L, keyword("storage", Monitored))
  EvGet(E_EndCollect)
  L2 := [ ]
  every put(L2, keyword("storage", Monitored))
  write("reclaimed ",integer(real(L[2] – L2[2]) / L[2] * 100), " percent of the string region")
  write("reclaimed ",integer(real(L[3] – L2[3]) / L[3] * 100), " percent of the block region")
  }
```

4.4 Monitoring String Scanning

As a descendant of SNOBOL4 and Icon, Unicon has a natural orientation towards text processing and includes a control structure devoted to that task. This section presents a brief overview of Unicon's string scanning facilities and then gives example execution monitors that portray the target program's use of this control structure. The examples are themselves relatively simple, but demonstrate the framework's capabilities in this area and are suggestive of more advanced possibilities to be explored in this domain using the framework. Techniques for monitoring string scanning can be built by extending the techniques presented for monitoring procedure and operator activity in Sect. 4.2.

4.4.1 Overview of String Scanning

Unicon's string scanning facility provides high-level text processing capabilities that free the programmer to think in terms of patterns in the text instead of character-by-character handling of indices and subscripts. String scanning operations work within the context of a string being scanned, the *subject*, and a current *position* of interest within that subject. Together, the subject and position form a *scanning environment* (Fig. 4.15).

The Unicon expression

```
s ? expr
```

subject "the yellow brick road"

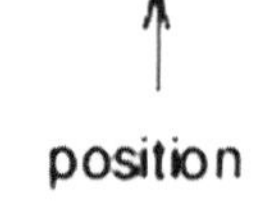

Fig. 4.15 A string scanning environment

evaluates *expr* in a scanning environment that consists of subject `s` and an initial position of 1 (the beginning of the string). Scanning environments remain in effect inside any procedure calls within *expr*. Scanning environments may be nested; the outer scanning environment is saved and restored when the inner environment is entered and exited.

Operations on scanning environments include absolute and relative movement of the position as well as various forms of string and character set matching and searching. Relatively sophisticated parsing is performed by using these operators in conjunction with goal-directed evaluation and backtracking. In particular, the functions that change position within an environment, `move()` and `tab()`, undo their effects if they are resumed by backtracking.

4.4.2 String Scanning Events

Since a TP may suspend from and later resume a scanning environment, string scanning instrumentation includes a set of events for environment creation, suspension, resumption, failure, and removal, analogous to the events that occur as a result of procedure activity. Monitoring string scanning may entail the maintenance of a scanning environment tree using code similar to the procedure activity tree presented in Sect. 4.2.

In addition to these events, string scanning position changes result in the occurrence of `E_Spos` events. If the scanning position is restored by `move()` or `tab()` during backtracking, a second `E_Spos` event occurs.

Scanning environment activity including position change events can be selected by an EM using the library symbol `ScanMask` as the argument to `EvGet()`. In addition to `ScanMask` events, a string scanning monitor may be interested in calls to the built-in string scanning functions that comprise Unicon's string analysis primitives, such as `find()` and `upto()`.

4.4.3 Absolute and Relative Position Changes

This section gives two simple EMs that present position change information with different emphases: (1) a view that portrays absolute position, and (2) a view that emphasizes relative position changes.

4.4.3.1 Visualizing Absolute Positions Within the Subject

String scanning operations move the position of interest within the subject forward or backward. Moving the position forward is common; moving the position backward is less common and usually is triggered by backtracking during goal-directed evaluation. It is useful to be able to observe when the position moves forward or backward and how large the changes in position are relative to the size of the string.

The following program displays an animated strip chart with subject lengths and position change information. For each position change event, the length of the subject is drawn down from the top and filled with two or three colors: a red segment indicates the current position or the number of characters already processed, while a white segment indicates the remainder of the string not yet processed. If backtracking has occurred, a gray segment in between the red and the white indicates the furthest forward that the scanning position has reached or the extent of the backtracking. A sample screen image is given in Fig. 4.16.

The program starts with standard initialization code, including the creation of window bindings for drawing segments in red and gray. The width of each bar is determined by variable `barwidth`, and the number of pixels drawn per character in the various segments is specified in the variable `scale`.

The program's main loop requests position change events, and plots a segment on the window for each change. `DrawRectangle()` draws a black outline to indicate the size of the scanned subject; calls to `FillRectangle()` plot the red and gray segments. A variable `max` holds the furthest position reached during scanning of a particular subject string; the gray segment is only drawn if backtracking has moved the position backwards into parts of the subject that have already been scanned.

```
while EvGet(E_Spos) do {
  s := keyword("subject", Monitored)
  position := &eventvalue
  if s == s_old then max <:= position
  else max := 1
  if *s > 0 then {
    DrawRectangle(x, 0, barwidth, scale * *s)
    FillRectangle(red, x, 0, barwidth, scale * (position – 1))
    if max > position then
      FillRectangle(gray, x, scale * (position – 1), barwidth, scale * (max – position))
    }
  x := (x + barwidth + 1) % WWidth()
  EraseArea((x + barwidth + 6) % WWidth(), 0, barwidth + 6)
  s_old := s
  }
```

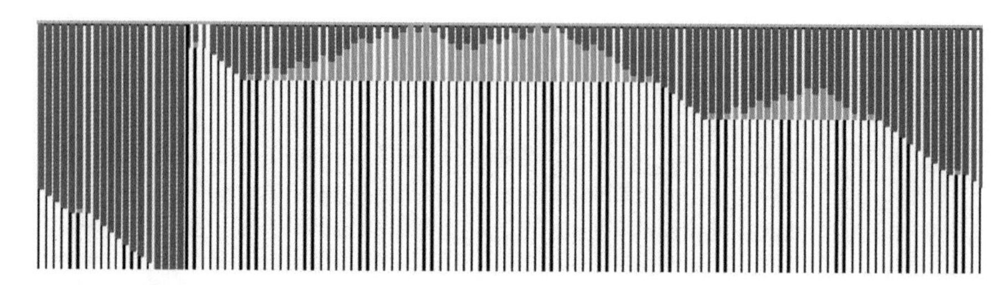

Fig. 4.16 Absolute string position

This simple EM does not scale its output to fit the window; in the event a very long subject is scanned, output is clipped to window boundaries. An additional limitation is that backtracking information is not saved and restored for nested scanning environments.

4.4.3.2 Visualizing Relative Position Changes

By tracking relative position changes, backward motion is highlighted and large position changes are emphasized. The following EM plots relative position change as distance from the middle of the window, with forward position change going below the midpoint and backward position change going up from the midpoint. A sample screen image is shown in Fig. 4.17.

After initialization, the main loop reads `E_Spos` events and uses the `keyword()` function to obtain the corresponding subject. If the subject is unchanged since the last event, the relative position change is noted. Like the previous example, this tool would provide more accurate information if it saved and restored the subject for nested scanning environments. The next section provides a method for doing so.

```
barwidth := 3
&window := open("pos", "g") | stop("can't open window")
x := 0
while EvGet(E_Spos) do {
  s := keyword("subject", Monitored)
  p := &eventvalue
  FillRectangle(x, WHeight() / 2, barwidth, 1)
  if s === s_old then
    if p > p_old then
      FillRectangle(x, WHeight() / 2, barwidth, p - p_old)
    else if p_old > p then
      FillRectangle(x, WHeight() / 2 - (p_old - p), barwidth, p_old - p)
  x := (x + barwidth + 1) % WWidth()
  EraseArea((x + barwidth + 6) % WWidth(), 0, barwidth + 6)
  s_old := s
  p_old := p
  }
```

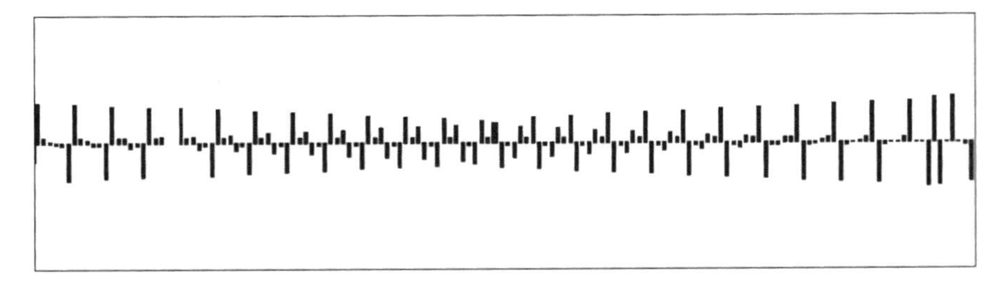

Fig. 4.17 Relative string position

4.4.4 The Scanning Environment Tree

Since scanning environments may be nested in much the same way as procedures, functions, and operators, it makes sense to use a tool similar to the Algae tool presented in Sect. 4.2 to portray nested scanning environments. String scanning function and operator activity can be animated within the hexagons allocated by Algae for each scanning environment. This provides a detail view of scanning within the surrounding program execution context.

A modified version of Algae that displays string functions and operators encoded as colors is shown in Fig. 4.18. The program uses the pinwheel metaphor from Sect. 4.3 to animate the sequence of operations independently within each scanning environment. Around the pinwheels' outside borders, circles are drawn in red, white, and gray segments to show current position and positional backtracking, similar to the absolute string positions example given earlier. The border around the pinwheel in the second column of Fig. 4.4 is almost entirely dark (the grayscale depiction of red), indicating that the scanning position is almost to the end of the string, while the border around the pinwheel in the fourth column is only slightly dark above the three o'clock position, showing that the scanning position is still near the front of the scanned string.

In order to add this kind of detailed information about string scanning environments, extra fields are added to Algae's activation record type for the current scanning position, the farthest scanning position reached in the scanning environment, and the environment's pinwheel angle (expressed in units of 1/64th of a degree).

```
record activation(node, parent, children, row, column, color, pos, maxpos, angle)
```

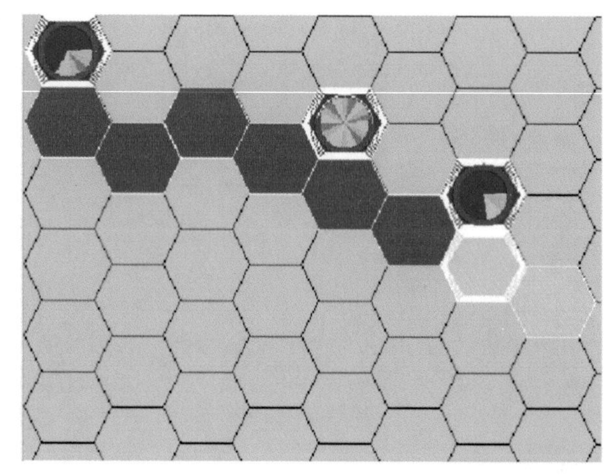

Fig. 4.18 Scanning environment trees and operations

4.4.4.1 Updating Position in the Current Scanning Environment

Position change events are added to the event mask passed to `evaltree()`. The case expression of the callback procedure for `E_Spos` events updates the current scanning environment's position fields, and draws red and gray arcs around the outside of the hexagon to show position information. Global variables `HexWidth` and `HexHeight` are used to determine the region inside the hexagon that is available for drawing.

Note that a callback static variable `scanenv` is used rather than the current activation (`new`), which can be a procedure, function, or operator called within the current scanning environment. Variable `scanenv` is maintained by code added to the case expression branches of Algae's `evaltree()` callback procedure, described below.

```
case &eventvalue of {
  # ... other Algae case branches as given in Section 4.2
  E_Spos: {
    scanenv.pos := &eventvalue
    scanenv.maxpos <:= &eventvalue
    unit := fullcircle / *scanenv.node
    DrawArc(red, hexcolumn_x(scanenv.col) + 5, hexrow_y(scanenv.row, scanenv.col) + 5,
          HexWidth - 10, HexHeight - 10, 0, (&eventvalue - 1) * unit)
    if scanenv.maxpos > scanenv.pos then
      DrawArc(gray, hexcolumn_x(scanenv.col) + 5, hexrow_y(scanenv.row, scanenv.col) + 5,
            HexWidth - 10, HexHeight - 10, (&eventvalue - 1) * unit,
            (scanenv.maxpos - scanenv.pos) * unit)
    DrawArc(wwhite, hexcolumn_x(scanenv.col) + 5, hexrow_y(scanenv.row, scanenv.col) + 5,
          HexWidth - 10, HexHeight - 10, (scanenv.maxpos - 1) * unit,
          fullcircle - (scanenv.maxpos - 1) * unit)
    }
  }
```

4.4.4.2 Drawing Pinwheel Sectors for Scanning Functions

The global table of colors is extended to map important string scanning functions onto window bindings with foreground colors that indicate which function is being performed. Activity that involves these functions is captured by adding code to the callback procedure's case expressions. The code for suspension events is shown here; similar code is added to the other cases.

```
!SuspCodes: {
  pinwheel(scanenv, \ Colors[new.node])
  # ... rest of code for suspension events
  }
```

4.4.4.3 Pinwheels for Nested Scanning Environments

The added fields of an activation record are initialized whenever a new scanning environment event is received. The modified code looks like:

```
!CallCodes: {
  # ... code as given in Section 4.2
  if &eventcode === E_Snew then {
    new.pos := new.maxpos := 1
    new.angle := 0
    }
  }
```

The pinwheel drawing procedure from Sect. 4.3 is revised to take an activation record and a window binding with a foreground color to encode the string operation being performed, and draw a single sector in that foreground each time it is called.

```
procedure pinwheel(arecord, win)
static full_circle, sector_units
initial {
  full_circle := 360 * 64
  sector_units := full_circle / 16 # 16 sectors in the circle
  }
  radians := −dtor(arecord.angle / 64)
  x := hexcolumn_x(arecord.col) + 6
  y := hexrow_y(arecord.row, arecord.col) + 6
  width := HexWidth − 12
  height := HexHeight − 12
  center_x := x + width / 2
  center_y := y + height / 2
  FillArc(arecord.color, x, y, width, height, arecord.angle + sector_units, blank)
  FillArc(win, x, y, width, height, arecord.angle, sector_units)
  DrawLine(arecord.color, center_x, center_y, radius * cos(radians) + center_x,
           radius * sin(radians) + center_y)
  arecord.angle +:= sector_units
  arecord.angle %:= full_circle
end
```

String scanning is an important feature in Unicon. In order to monitor it correctly, an EM must handle position changes as well as nested and suspended scanning environments. The extra attention required to monitor scanning correctly parallels the effort required to implement scanning correctly in the language.

Although string scanning is important, most programs use string scanning in extremely simple ways. Although detailed views will always be useful in debugging situations, in more general program understanding efforts, the information provided by literal text-oriented views of string scanning may be less useful than might be expected. A better approach may be to view string scanning within a larger context of program operation, such as the modified Algae example. It is not clear how to best monitor and visualize string scanning; this is still an open area for research.

4.5 Monitoring Structure and Variable Usage

Previous sections have demonstrated techniques for monitoring various aspects of program control and memory usage. Although some aspects of TP data usage are observable by means of memory allocation and garbage collection events, key aspects of program behavior are often characterized in terms of operations on program data, such as manipulations of program data structures or variable references.

This section presents techniques for monitoring data from both program-wide and narrower, variable-oriented viewpoints. Example EMs include list access monitors that show usage of Unicon's built-in list data type on a program-wide scale, and variable reference monitors that show activity within individual procedure activations. There are many other ways to present data structure activity, and this is an open area of research. The examples in this section illustrate the capabilities and possible uses of the framework in this domain.

4.5.1 Visualizing Lists and List Accesses

On a program-wide scale, a tool that visualizes list activity is representative of techniques needed to monitor list, table, record, set, and object types. The list data type is used for a variety of purposes. Some programs use a few large lists, while other programs may use hundreds or thousands of small lists. Lists can change in size dynamically using both queue and stack operations, and they can also be accessed randomly similar to arrays in other languages.

The following EM portrays an overall view of list behavior, portraying TP lists as a sequence of vertical bars with length proportional to the size of the list. Vertical segments of the bars are color-coded by the types of the list elements. If a list's elements are of the same type, this forms a solid bar of that type's color; if a list is heterogeneous, its appearance is candy striped with the various colors of its elements' types. The horizontal position of a list's bar on the display is given by the list's serial number. A serial number is an integer associated with each list when it is created. Using serial numbers to determine screen position orders the lists from left to right by time of creation.

Queue, stack, and array-style random accesses are portrayed by changing the size of the bar (in the case of queue and stack accesses) or briefly painting a segment of the bar black and then redrawing it (in the case of random accesses). An example image from this program is given in Fig. 4.19. Empty columns in this view indicate

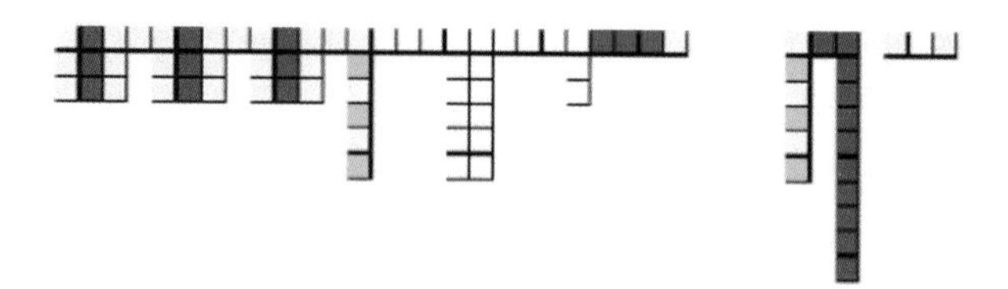

Fig. 4.19 A list access monitor

serial numbers at which no list has yet been created (on the far right) or lists that are empty or have been garbage collected (in the middle of the figure).

One of the key features of this program is a degree of scalability necessary in order to accommodate programs with large numbers of lists and yet present as much detail as screen space allows. In particular, if the number of lists is too large to fit in the window, the window is split into two rows and the number of vertical pixels per element is halved; this generalizes to *n* rows of as few as one vertical pixel per list element. Figure 4.20 depicts a view in which the number of lists has caused a split into two rows.

Figure 4.21 depicts a scaled image for a larger number (around 400) of lists requiring eight rows. Spaces in the figures again generally indicate empty or garbage-collected lists. Despite some effort expended to achieve scalability, there are limits of scale in for any literal depiction of individual elements within all the lists in a program execution.

Scalability to hundreds of lists is achieved by maintaining a number of interdependent variables to describe the screen geometry. The window is divided into a matrix of size `rows` by `cols` corresponding to individual lists; each element of the matrix is in turn divided into vertical segments of height `elem_height`.

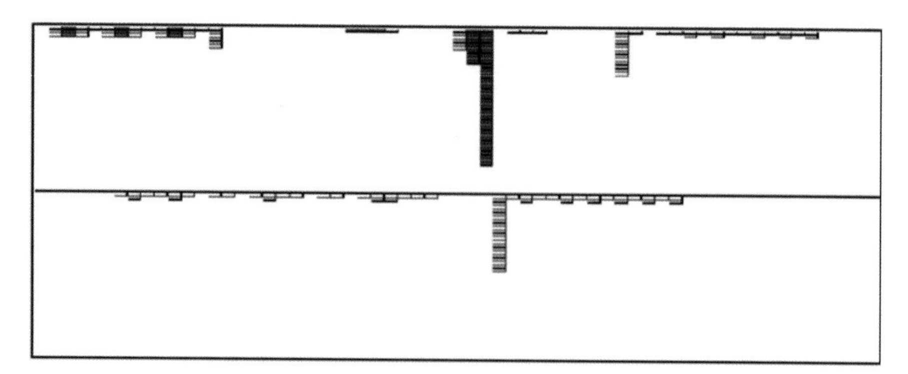

Fig. 4.20 A moderate number of lists

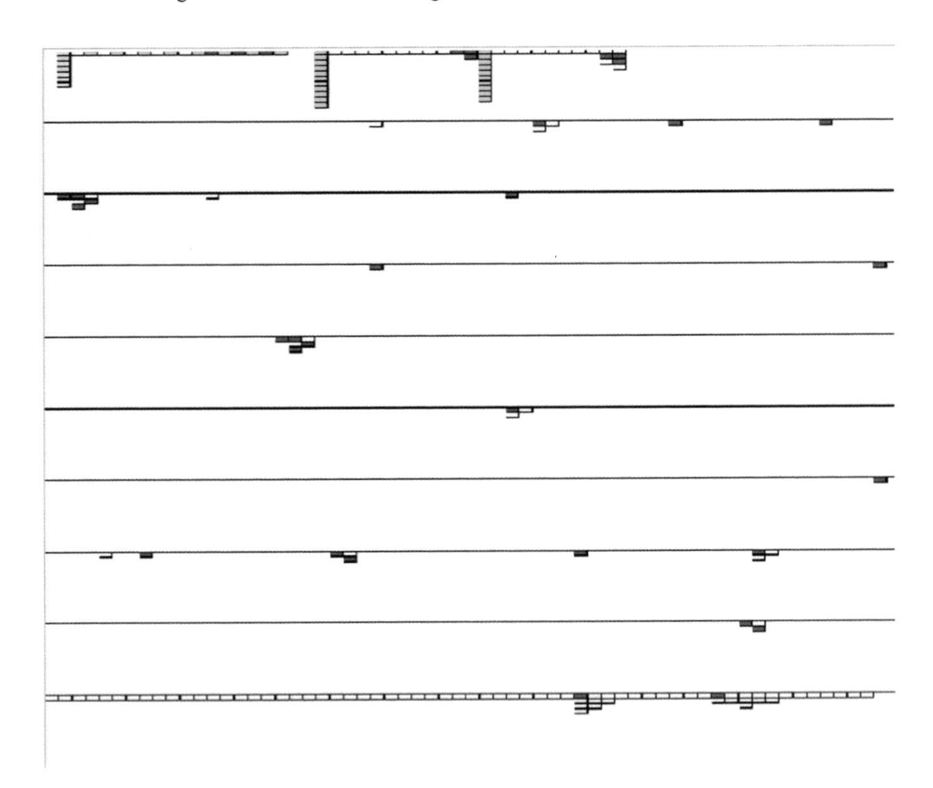

Fig. 4.21 A large number of lists

```
global rows,       # number of rows of entire lists
    cols,          # number of lists displayed per row
    elem_height    # height of an individual list element
```

In addition to this basic screen geometry, a count of the number of lists in TP is kept in `number_active`, and the mapping from lists to window (row, column) coordinates is maintained in table `list_locations`. The mapping from lists to window coordinates uses list serial numbers as keys, rather than list values themselves. If the EM retained references to the TP lists instead of their serial numbers, none of the TP lists could be reclaimed by garbage collection.

Procedure `redraw()` draws an entire picture of all the lists in the program. It uses the Unicon function `structure()` to generate all the allocated structures in the program, and assigns each list a row and column. Each element of each list is then drawn by `FillRectangle()` in a color determined by the element's type by a call to `objcolor()`.

```
procedure redraw()
  EraseArea()
  column_width := WWidth() / cols
  row_height := WHeight() / rows
  every i := 1 to rows — 1 do
    DrawLine(0, i * row_height, WWidth(), i * row_height)
  number_active := 0
  list_locations := table()
  every type(L := structure(Monitored)) == "list" do { # for every list
    number_active +:= 1
    row := 1 + number_active / cols
    col := number_active % cols
    list_locations[serial(L)] := location(row, col)
    every index := 1 to *L do
      Fill Rectangle(objcolor(L[index]), col * column_width, (row — 1) *
          row_height +(index — 1) * elem_height + 1, column_width, elem_height)
    }
end
```

Procedure `redraw()` is called whenever the scaling must be changed. The view it establishes can be updated incrementally for ordinary list construction and access by drawing one or more individual list elements with procedure `plot()`. `plot()` draws a rectangle, first with a black rectangle to highlight the access, and then with a rectangle of a specified color.

```
procedure plot(w, row, col, index, del)
  /del := 40
  x := col * column_width
  y := (row — 1) * row_height + (index — 1) * elem_height + 1
  if del > 0 then {
    FillRectangle(vblack, x, y, column_width, elem_height)
    WFlush(vblack)
    delay(del)
    }
  FillRectangle(w, x, y, column_width, elem_height)
end
```

The main loop fetches list events and updates by calling `plot()`. `redraw()` is called when the screen becomes full or the window size changes. One significant detail of list access monitoring is that a list access results in two events, one (`E_Lref`) with the list itself for an event value, and a second event (`E_Lsub`) with an integer event value that gives the index accessed within the list. EM saves the list value in the first event and uses it when the second is reported. Since the events come in pairs, TP does not do anything in between the two events, but after the second event, EM must use and then destroy its reference to the list or it might spuriously prevent the list from being garbage-collected.

```
while EvGet(ListMask) do
  case &eventcode of {
    E_Lref : L := &eventvalue
    E_Lsub : {
      index := &eventvalue
      if index < 0 then
        index +:= *L + 1
      loc := list_locations[serial(L)]
      plot(objcolor(L[index]), vertical(loc), horizontal(loc), index)
      L := &null
      }
    # ... other events handled similarly
    }
```

Although this example uses some sophistication to scale well to larger numbers of lists, it can be enhanced in various ways. For example, relaxing the direct mapping from serial number to screen location would allow screen space to be reclaimed whenever a list was garbage-collected. Another improvement would be to portray list operations in a visually distinct way instead of simply maintaining an accurate representation of the lists' contents.

4.5.2 Visualizing Heterogeneous Structures

Visualization poses challenges when structure types are treated in isolation, as in the preceding section. The challenges become harder when relationships between interconnected structures of different types are considered. Advanced languages such as Unicon support *heterogeneous* structures: not just lists of strings and lists of lists, but lists of arbitrary elements that may be strings, lists, integers...whatever. For example, a hash table might implement both directions of a two-way table between a set of string names and integer values, containing keys of each type that map to values of the other type. When structures are heterogeneous, it is productive to consider visualizing them all together in the same image. In this case, the different methods of accessing data are what distinguish different data types. Since a single type may implement different interfaces with different access methods, this heterogeneity may actually be present when visualizing a single type. Unicon lists, for example, offer array, stack, and queue access.

Both heterogeneous data values and heterogeneous access methods pose problems in debugging and in visualization. Textually naming elements several levels deep in a multi-level structure is tedious, and heterogeneous values only exacerbate the problem since the name used to access a sublevel may not be known until the type of the upper level is queried. More generally, characterizing the conditions of interest and distinguishing them from expected cases can be difficult or tedious, especially if the structure is complex.

The monitor described in this section is a heterogeneous structure visualizer, affectionately called the Structure Spy. It was written by Kevin Templer and Wenyi Zhou. This program displays relationships between arbitrary interconnected structures such as link lists, trees, and graphs. Nodes may be built from any Unicon structure type, including lists, tables, sets, and records, which are color-coded. The Structure Spy is concerned only with interesting multilevel structures; structures that contain only references to nonstructures are ignored.

From the Spy's point of view, all structures are equivalent and the operations the Spy looks for are node creation and destruction, intracomponent changes, and splits and merges of connected components. Although structure references are unidirectional in Unicon, all references are treated as bidirectional when determining the connected components.

As the target program runs and builds up structures, each connected component is depicted in a separate area in the window. Components' areas are subdivided into quarters and reallocated whenever more space is needed. Programs with a few large components look very different from programs with many small components. Figure 4.22 depicts a sample image of a single component consisting of interconnected nodes from a mixture of list and record data types.

Merges of components happen when one large structure is inserted or linked to another. When this happens, the components being merged are briefly highlighted with light hashed lines, as shown in Fig. 4.23 (left).

After a merger, the combined component is drawn in one of the two areas that were merged, and the other area is vacant until a new component is created and placed there. Figure 4.23 (right) shows the window after the merger is complete.

The source code for the Structure Spy is too lengthy to present in its complete form; we instead compare its implementation to the list monitor described above. The `main()` procedure uses a mask that asks for the union of the predefined sets of structure events, the assignment events, and the garbage-collection events. A table named `Handlers` maps event codes to associated procedures. For each event, the corresponding handler procedure is called via the table.

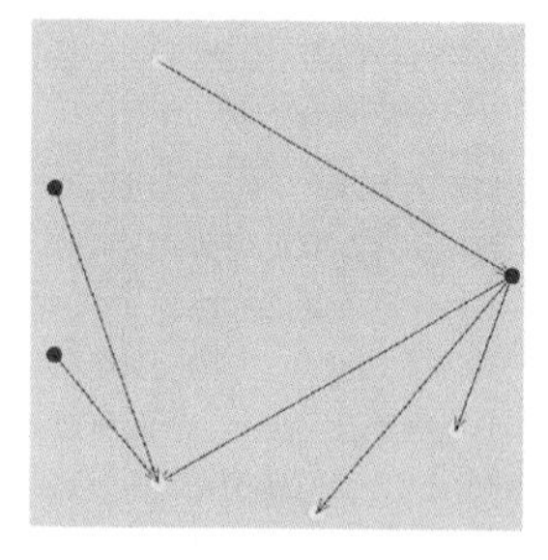

Fig. 4.22 Heterogeneous structure visualization

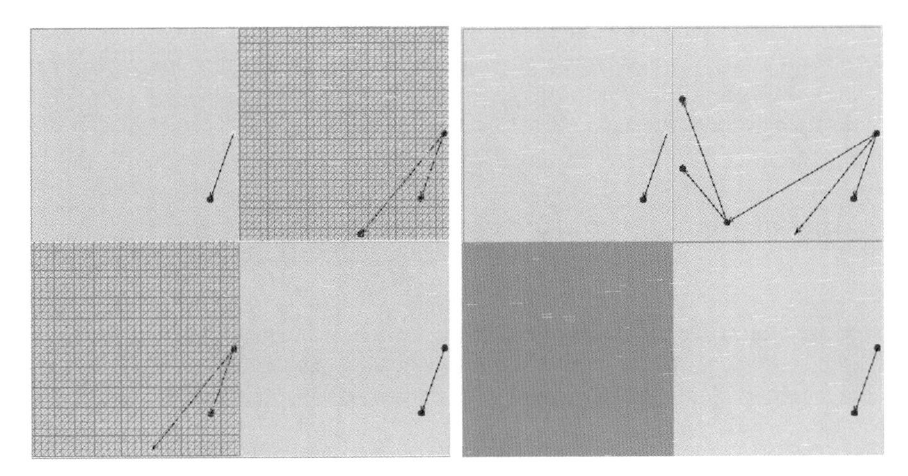

Fig. 4.23 Before and after a merge of two components

```
# initialization, including EvInit()
mask := StructMask ++ AssignMask ++ E_Collect ++ E_EndCollect
while EvGet(mask) {
  # ...
  ( \ (Handlers[&eventcode])) ()
  }
```

The event handler procedures examine the event value. If the structure operation is on a multilevel structure, its effect on the structure is classified (add a link, delete a link, etc). The handler procedure for the list `put()` function is given below. The event value after an `E_Lput` event is the list after the element has been added to the end; the newly added element on the end is thus `&eventvalue[-1]`.

`AddObjLink()` inserts the new link into the Spy's model of the program structures; its parameters are source node, index or key, and destination node. `ObjId` is a function that computes a string containing the type and serial number of the event value for each event; these strings represent the object as a key in various internal tables.

```
procedure ListPut()
  if IsObj(type(&eventvalue[-1]), 1) then {
    AddObjLink(ObjId(&eventvalue), *&eventvalue, ObjId(&eventvalue[-1]))
    }
end
```

The source code for the graphical layout is too long to present in detail here. The layout is kept simple and is geared for scalably animating arbitrary programs' behavior, rather than presenting attractive still shots of special cases. Components are drawn in squares, and these squares are subdivided evenly into four pieces when more space is needed for new components. The layout of an individual component simply maps nodes around a circle to preserve the visibility of edges.

4.5.3 Monitoring Variable References

Monitoring structure accesses with techniques such as those described in the previous section is useful, but in many EMs, notably debuggers, data monitoring is driven from the variables used in the program. We consider two examples of variable monitoring, one that visualizes all variables and one that identifies references to specific variables of interest.

4.5.3.1 Assignment Events

One of the most common monitoring techniques is the observation of assignments, where the user is informed or monitoring code is executed whenever an assignment to a particular variable or set of variables is made. The instrumentation reports an `E_Assign` event on each assignment. `E_Assign` has a string event value equivalent to calling `name(v)` on the assigned variable, suffixed by a *scope code*. The scope codes are listed in Table 4.1.

Statics, locals, and parameters are followed by the name of the procedure in which they are defined. For example, a local variable `i` in procedure `main()` would produce an `E_Assign` event value `"i-main"`. Variable references to structure elements have no scope code.

For assignments to named variables and keywords, the name and scope are sufficient to perform reference detection; the name and scope may be augmented by procedure activity information in order to provide finer detail for local (and especially recursive local) variables. For assignments to structure elements, the event value cannot produce the name. A given structure element might be assigned by means of any of several variables that reference the structure. For this reason, reference detection techniques are different for named variables and for structure element variables.

Table 4.1 Scope Codes

Code	Scope
"+"	Global
":"	Static
"-"	Local
"^"	Parameter

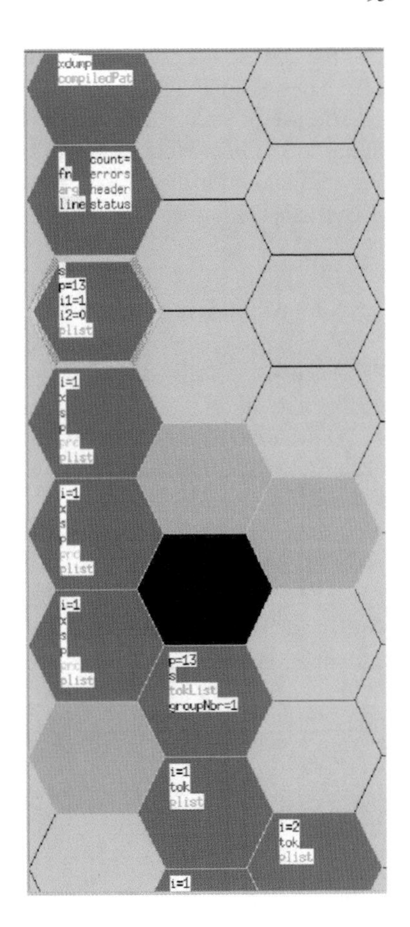

Fig. 4.24 Monitoring variables in active procedures

4.5.3.2 Monitoring Variables by Name

Figure 4.24 shows a window image of a tool that displays the names and types of variables associated with procedure activations; the names are written in multiple columns in the case of a procedure with a larger number of variables. As its appearance indicates, the tool is an enlarged version of the Algae program. The names of procedure parameters and local variables are displayed within each activation, drawn in a color that indicates the type of the variable. Colors are updated after each assignment. One useful extension to this tool is to show the values of integers. This is useful because integers are common, because they do not require much space, and because they are not heap-allocated and therefore do not appear in other data-oriented monitors.

The required modifications to Algae source code are omitted here for the sake of brevity; they are comparable to the extensions for string scanning given in the preceding chapter. The technique used is the monitoring of assignment events, considering only those events whose scope code indicates either a local variable or parameter assignment.

The use of source text names creates serious spatial problems. Another reasonable way to extend this EM would be to modify it to use smaller rectangles for each variable and omit the names. Specific variables' names could be shown when the user clicks the mouse atop a particular variable.

4.5.3.3 Monitoring Individual Variables

A named variable is identified by its name and scope, or by its instantiating procedure activation if recursively created local variables are considered distinct. For such variables, reference detection is implemented using the `E_Assign` event values and some additional logic. Two examples below illustrate cases where (1) the EM acts on any assignment to a variable defined within a given procedure, and (2) the EM acts on assignments only within a specific activation record.

In the nonrecursive case, variables can be identified by their name and scope. A collection of variable names of interest might be stored in a Unicon set ("trapped_variables" in the code below). Variable traps require selection of assignment events and maintenance of current procedure information using the `evaltree()` procedure, as described in Sect. 4.2 on following procedure activity. The correct invocation of `evaltree()` is:

```
evaltree(ProcMask ++ E_Assign, trap_callback, activation_record)
```

Procedure `trap_callback()` detects variable references with a set membership test.

```
procedure trap_callback(current_proc)
  if &eventcode === E_Assign then
    if member(trapped_variables, &eventvalue) then {
      # perform trap
      }
end
```

In some EM's, the handling of recursive procedure calls requires a more sophisticated form of variable trapping in which each individual local variable within each procedure activation record is treated as a distinct entity and can be trapped separately. This is relevant in recursive procedure calls. This form of trapping can be implemented by adding a field to the structure maintained for activation records:

```
record trapped_activation(p, parent, children, trapped_variables)
```

The variable reference detection is performed using this record type in an `evaltree()` invocation of the form given above, replacing the line

```
if member(trapped_variables, &eventvalue) then {
```

in `trap_callback()` with the line

```
if member(new.trapped_variables, &eventvalue) then {
```

4.5.3.4 Detecting Structure Variable References

Unicon structures have pointer semantics. Consequently, if two variables refer to the same structure, a trap on the name of an element of one of the variables will not catch an assignment using the other variable name. In the code

```
L1 := list(2)
L2 := L1
L2[1] := "foo"
```

a trap on variable `L1[1]` will not catch the assignment even though assignment is made to it. In order to trap structure elements, the information provided in assignment events need to be mapped down to the underlying structure.

Unfortunately, `name(v)` for a structure variable produces only a type code letter and a string image of the subscripting element. Without resorting to data-intrusive techniques such as altering the internal representation of structures, monitors cannot tell from an assignment to an element which structure the element is in. Instead, monitors use the framework's extensive access to the program state.

Given the information `E_Assign` events provide about structure assignments, one way to trap structure elements is to check if a structure assignment *might* be a variable trap, and then compare all structures that might have been changed, after the assignment has been performed. In general, nonintrusive techniques for monitoring assignments are inefficient: this particular approach imposes a cost on structure variable assignments proportional to the number of trapped structure variables of the same type and index; if a large number of variables are to be trapped, data-intrusive techniques may be needed for performance reasons. An appropriate trapped variable technique has been developed for SNOBOL4 [2]. For every trapped structure variable, a triple consisting of the structure, the index or key, and the old value is maintained.

```
record trapped_structvar(struct, index, value)
```

These records are stored in a table, indexed by the string name that is reported by `E_Assign` when the variable is assigned.

Structure variable traps use not only `E_Assign` events, but also the `E_Value` events that are produced following the assignment. If the structure indexed by the key does not still equal the old value, the assignment has taken place. This technique is not capable of detecting assignments of the same value replacing itself in structures. The code is

```
codes := E_Assign ++ E_Value
  while EvGet(codes) do
    case &eventcode of {
      E_Assign : {
          if match("T[" | "L[" | "R.", &eventvalue) then
            struct_asgn := trapped_structs[ &eventvalue ]
          else struct_asgn := &null
          }
        E_Value : {
          every tv := ! \ struct_asgn do
            if tv.struct [ tv.index ] ~=== tv.value then {
              # the trapped structure element has been assigned
              }
          }
        }
```

This technique works directly for tables and lists. It also works for record fields as long as the field is translated into its corresponding index for insertion into the `trapped_structvar` record.

References

1. R. E. Griswold, “Views of Storage Allocation in Icon,” 1992.
2. D. R. Hanson, “Event Associations in SNOBOL4 for Program Debugging,” *Software—Practice and Experience*, vol. 8, pp. 115–129, 1978.

Chapter 5
Integrating Multiple Views

This chapter covers:

- Monitor coordinators
- Integration within a 2D window via subwindows

As illustrated in the preceding chapters, the Unicon execution monitoring interface makes it easy to develop new EMs. In this model, monitors are free to specialize in particular aspects of program execution, and the user selects the aspects to monitor in a given execution. When multiple EMs come into play, the selection of which EMs to use, the execution of those EMs, and their communication interface are the responsibility of a program called a monitor coordinator (MC).

Monitor coordinators are a central unifying aspect of this book in that they are the abstraction that we embed into a virtual environment in order to integrate software visualizations within that environment. After a general discussion of monitor coordinators, an example monitor coordinator is presented that implements a generalization of the *selective broadcast* communication paradigm advocated by Reiss [1]. Other paradigms of monitor coordination are possible within the framework. In addition, other generalizations of selective broadcast proposed in the literature may prove complementary to the one presented in this chapter [2].

5.1 Monitor Coordinators

Unicon execution events are always reported to the parent program that loaded the TP being monitored. This means that the normal event reporting mechanism handles simple relationships such as monitoring a monitor or monitoring multiple TPs (Fig. 5.1).

On the other hand, the parental event report relationship means that if more than one EM is to monitor a TP, the TP's parent must provide other EMs with artificial copies of the TP events; Unicon's `EvSend()` function provides this service. Figure 5.2 depicts a parent EM that forwards TP events to an assisting EM.

C. Jeffery, J. Al-Gharaibeh, *Writing Virtual Environments for Software Visualization*, DOI 10.1007/978-1-4614-1755-2_5

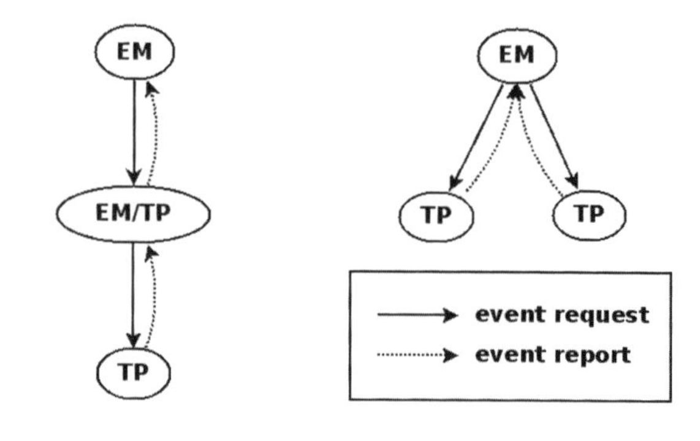

Fig. 5.1 Monitoring a monitor; monitoring multiple TPs

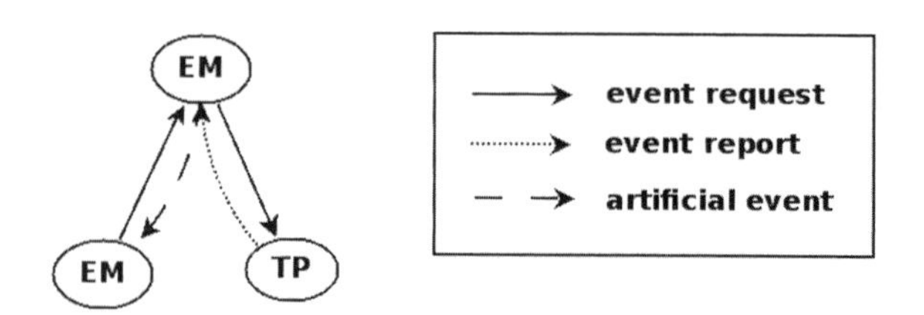

Fig. 5.2 Forwarding events to an assistant

Monitor coordinators are specialized EMs whose primary function is to forward events to other client EMs. A monitor coordinator is an event monitoring *kernel* that integrates and coordinates the operation of multiple standalone tools. By analogy to operating systems, the alternative to a kernel design would be a monolithic program execution monitor that integrates all operations into a single program.

Figure 5.3 depicts some relationships among MCs. Figure 5.3a is similar to Fig. 5.2 and shows that an MC is just an execution monitor that forwards events. Figure 5.3b shows the main purpose for MCs, the execution of multiple EMs on a single TP. Figure 5.3c shows an MC monitoring an MC.

MC configurations and logic are generally limited to and revolve around parent-child relationships. For example, it is impossible to monitor events in a TP loaded and being monitored by another EM or MC unless that parent is configured to forward such events.

Since event reports also transfer control, MCs are also schedulers for EMs, relinquishing the CPU to them by forwarding events to them. In the simplest case, the MC forwards an event and waits for the EM to request another event before continuing; this scheduling is a form of cooperative multitasking. If the MC is the parent that loaded the EM in question, it can request event reports (such as clock ticks) from the EM in order to preempt its execution. Since MCs are special-purpose EMs, development of an efficient MC design falls within the scope of exploratory programming support provided by Unicon.

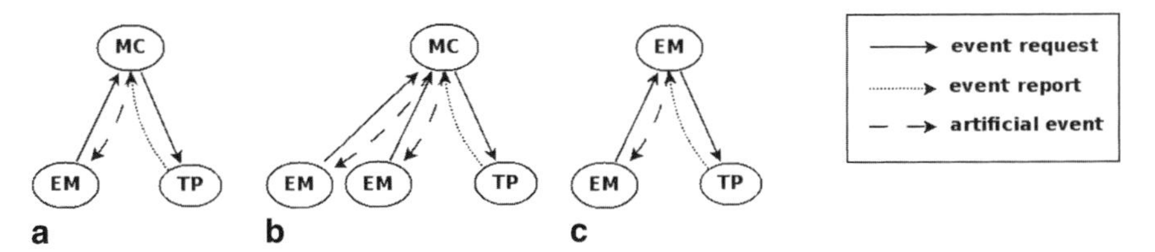

Fig. 5.3 Monitor coordinators

Monitor coordinators have several desirable attributes. With an MC, monitors can be developed independently of one another and of the MC itself; they can run as standalone monitors, directly loading and executing the program to be monitored. This allows monitors to be debugged separately and puts firewalls between monitors when they monitor the same program at the same time. Support for multiple monitors allows EMs to be written to observe very specific program behavior and still be used in a more general setting. This, in turn, reduces the burden of generality placed on EM authors. Specialization also simplifies the task of presenting information, since each EM uses its own window and the user decides how much attention and screen space to devote to each EM. Also, monitor coordinators are extensible. It is easy to add new tools to the visualization environment. Adding a new tool to run under a MC does not require recompiling or even relinking the MC or any of the other visualization tools.

Monitor coordinators do have disadvantages. The implementation of MCs poses serious performance problems that require careful consideration. Although unsuitable for exploratory monitor development and experimental work, a single monolithic EM provides better performance than a MC that loads multiple EMs.

The primary problem with MCs is the number of context switches among tasks; on some architectures, switching between coroutines is an expensive operation. Minimizing the number of switches required must be a goal of most MC designs.

5.2 Eve: The Reference Monitor Coordinator

Eve is an example MC that allows the user to execute one or more EMs selected from a list and forwards TP events to those EMs that the user selects. The communication provided by Eve represents a generalization of the selective broadcast communications paradigm, because EMs may change the set of events at any time during execution; in Reiss's FIELD system, tools can specify the set of events they are interested in only when they are started. Unlike Forest's generalization of selective broadcast in which dynamic control is achieved by placing a greater computational load on the coordinating message server, Eve maintains an extremely simple message dispatch mechanism and passes policy changes on to the TP by recomputing the TP's event mask whenever needed. By suppressing events as early as possible, the higher performance required for execution monitoring is attained.

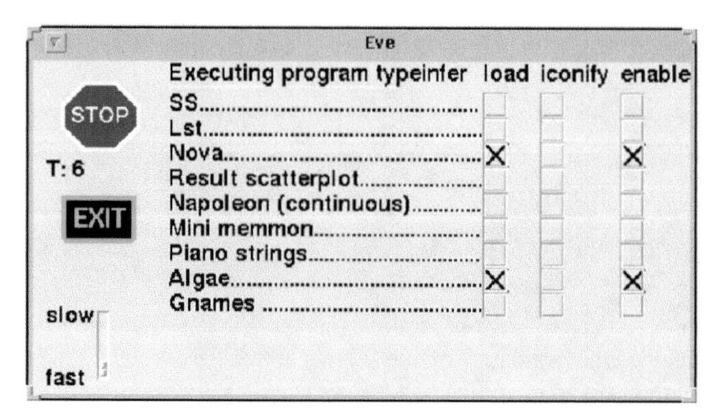

Fig. 5.4 Eve's control window

This technique of continually minimizing the set of events reported by the TP could be used in conjunction with a Forest-style policy mechanism in the monitor coordinator, if that were desired.

Eve is a cooperative multitasking scheduler. Figure 5.4 shows an image of Eve's control window. Eve has a very simple graphical user interface, in addition to the interfaces of the monitors that it runs. On the lefthand side are buttons that pause and terminate TP execution and a slider that controls execution speed. The main area of the window consists of a configurable list of EMs, and for each EM a set of buttons allow the tool to be controlled during TP execution. In the figure, two EMs are loaded and enabled.

Several techniques are employed in Eve to obtain good performance. The key ideas are to filter events at the source and to precompute the set of EMs to which each event code is distributed.

Different EMs require different kinds of events. After obtaining a list of client EMs to execute, Eve loads each client. It then activates each EM for the first time; when the EM completes its initialization, it calls `EvGet()`, passing Eve an event mask.

5.2.1 Computation of the Minimal Event Set

Each time an EM requests its next event report from Eve, it transmits a cset event mask indicating which events it is interested in. Eve could simply request all events from the TP, and forward event reports to each EM based on its current mask. The interpreter runtime system is instrumented with so many events that this brute-force approach is too slow in practice. In order to minimize the cost of monitoring, Eve asks the TP for the smallest set of events required to satisfy the EMs.

From the event masks of all EMs, Eve computes the union and uses this cset to specify events from the TP. The code for this union calculation is

```
unioncset := ''
every monitor := !clients do
  if monitor.enabled === E_Enable then
    unioncset ++:= monitor.mask
```

Every EM can potentially change its event mask every time it requests an event. Constant recomputation of the union mask would be unacceptably expensive. Fortunately, most tools call `EvGet()` with the same event mask cset over and over again. Eve does not recompute the union event mask unless an EM's event mask changes from the EM's preceding event request.

5.2.2 *The Event Code Table*

The minimal event set described above greatly reduces the number of events actually reported from the TP. When an event report is received from the TP, Eve dispatches the report to those EMs that requested events of that type. The larger the number of EMs running, and the more specialized the EMs are, the smaller the percentage of EMs that typically are interested in any given event.
Eve could simply test the event code with each EM's cset mask with a call `any(mask, &eventcode)`. This test is fast, but performing the test for each EM is inefficient when the number of EMs is large and the percentage of EMs interested in most events is small. Instead, the list of EMs interested in each event code is precomputed as the union mask is constructed. These lists are stored in a table indexed by the event code. Then, after each event is received, a single table lookup suffices to supply the list of interested EMs. For each enabled monitor, the code for union mask computation is augmented with:

```
every c := !monitor.mask do {
  /EventCodeTable[c] := [ ]
  put(EventCodeTable[c], monitor)
  }
```

5.2.3 *Event Handling*

Eve requests three types of events whether or not any client EM has requested them: `E_Tick`, `E_MXevent`, and `E_Error`. Eve uses these events to provide basic services while execution is taking place; since these events occur relatively infrequently, they do not impose a great deal of overhead.

`E_Tick` events allow Eve to maintain a simple execution clock on the control panel. `E_MXevent` events allow Eve to receive user input (such as a change in the slider that controls the rate of execution) in its control panel. `E_Error` events allow Eve to handle runtime errors in the TP and notify the user when they occur, allowing errors to be converted to expression failure at the user's discretion. As in the Structure Spy in Chap. 4, handlers for events are called through a global table indexed by event code, allowing for easy extension.

5.2.4 Eve's Main Loop

Eve's main loop activates the TP to obtain an event report, and then dispatches the report to each EM whose mask includes the event code. Since this loop is central to the performance of the overall system, it is coded carefully. Event dispatching to client EMs costs one table lookup plus a number of operations performed for each EM that is interested in the event—EMs for which an event is of no interest do not add processing time for that event. The code for Eve's main loop is:

```
while EvGet(unioncset) do {
 #
 # Call Eve's own handler for this event, if there is one.
 #
 (\ EveHandlers[&eventcode]) ()
 #
 # Forward the event to those EMs that want it.
 #
 every monitor := !EventCodeTable[&eventcode] do
   if C := EvSend( , , monitor.prog) then {
     if C ~=== monitor.mask then {
       while type(C) ~== "cset" do {
         #
         # The EM has raised a signal; pass it on, then
         # return to the client to get its next event request.
         #
         broadcast(C, monitor)
         if not (C ~= EvSend( , , monitor.prog)) then {
           unschedule(monitor)
           break next
           }
         }
       if monitor.mask ~===:= C then
         computeUnionMask()
         }
       }
   else
     unschedule(monitor)
   # if the slider is not zero, insert delay time
 }
```

5.2.5 *Interactive Error Conversion*

Normally, execution terminates when a runtime error occurs. Unicon supports a feature called *error conversion* that allows errors to be converted into expression failure. Error conversion can be turned on and off by the source program by assigning an integer to the keyword `&error`. `&error` indicates the number of errors to convert to failure before terminating the program; on each error the value of `&error` is decremented and if it reaches zero, the program terminates. A program can effectively specify that all errors should be converted by setting `&error` to a small negative integer. The mechanism is limited in that it does not allow the user or the program to inspect the situation and determine whether error conversion is appropriate: error conversion is either on or it is off.

Eve catches runtime errors in the TP and allows the user to decide whether to terminate execution or convert the error into expression failure and continue execution (Fig. 5.5).

An `E_Error` event occurs upon a runtime error. A monitor that requests `E_Error` events is given control before the error is resolved. Eve requests these events, presents the user with the error, and asks for an appropriate action. The code in Eve that does interactive error conversion is:

```
procedure eveError()
  win := open("Runtime error " || &eventvalue, "g")
  write(win, "Runtime error ", &eventvalue)
  write(win, "File ", keyword("file", Monitored), "; line ", keyword("line", Monitored))
  write(win, "errortext", Monitored))
  writes(win, "Convert to failure? ")
  if read(win) == "y" then
    keyword("error", Monitored) := 1
  close(win)
end
```

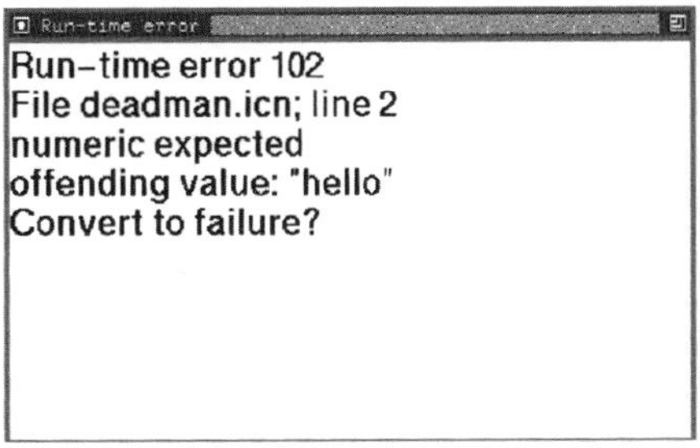

Fig. 5.5 Eve's interactive error converter

Run-time error
Run-time error 102
File deadman.icn; line 2
numeric expected
offending value: "hello"
Convert to failure?

5.3 Writing Monitors that Run Under a Monitor Coordinator

Eve supplies events to client EMs using the standard `EvGet()` interface described in Chap. 3. This section describes a few differences between the standalone interface and the Eve environment. Note that programs written for the Eve environment run without change or recompilation as standalone tools.

5.3.1 Client Environment

After each EM is loaded, Eve initializes it with references to its event source (the Eve program itself) and the TP. The former is necessary so that EMs yield control to Eve to obtain each event. The latter is provided so that the state of the TP may be examined or modified directly by all EMs. These references in the form of co-expression values are assigned to the keyword `&eventsource` and the global variable `Monitored`, respectively; the global variable `Monitored` is declared in each EM when it links to the `evinit` event monitoring library.

Under Eve, `&eventsource` is *not* the TP. For this reason EMs should always use `Monitored` to inspect program state. For example, to inspect the name of the current source file in the executing program, an EM should call `keyword("file", Monitored)` rather than `keyword("file", &eventsource)`.

Aside from the fact that `&eventsource` is not `Monitored` under Eve, from a programmer's standpoint, Eve's operation is implicit. Just as monitoring does not inherently affect TP behavior (other than slowing execution), within the various EMs Eve's presence normally is not visible; the EM can call `EvGet()` as usual.

5.3.2 General-Purpose Artificial Events

Eve sends certain artificial events when directed by the user (in the Eve control window). These include the disable and enable events discussed above, `E_Disable` and `E_Enable`. A tool can pass a second parameter to `EvGet()` in order to receive these pseudo events, for example `EvGet(mask, 1)`. When an `E_Disable` event is received, a tool is requested to disable itself. Tools that do not maintain any state between events can simply shut off their event stream by calling `EvGet(' ', 1)`:

```
case &eventcase of {
  # ... more frequent events come first
  E_Disable: while EvGet(' ', 1) ~=== E_Enable
  }
```

Tools that require events in order to maintain internal consistency might at least skip their window output operations while they are disabled. An `E_Enable` event informs the tool that it should resume operation, updating its display first if necessary.

5.3.3 Monitor Communication Example

In addition to the use of artificial events for communication between Eve and other EMs, artificial events can be used by EMs to communicate with each other, using Eve as an intermediary. For example, a line number monitor is more useful if the user can inquire about a section of interest in the line number graph and see the corresponding source text. This functionality can be built in to the line number monitor, but since many visualization tools can make use of such a service, it makes more sense to construct an EM to display source lines, and use virtual events to communicate requests for source code display from other EMs.

Communication using Eve starts with the definition of an artificial event code for use by the communicating EMs. Some of these codes such as `E_Disable` are defined in the standard library, but in general, EMs can use any artificial event codes that they agree upon. In this case, an event code, `E_ALoc,` is defined for artificial location display events. Communicating EMs also agree on the type and meaning of the associated event value. In this case, the associated event value is an integer encoding of a source line and column number, similar to that produced by `E_Loc` events.

The source code display EM is similar to other EMs, except that it is not interested in TP events, but only in `E_ALoc` events. Its main loop is

```
while EvGet('', 1) do
  if &eventcode === E_ALoc then {
    # process requests for source code display
    }
```

Any EM that wishes to request source location display services sends an `E_ALoc` event to Eve. Eve then broadcasts this event to those tools that requested artificial event reports. The code to send a location request event to Eve from within an EM is

```
loc := location(line, column)
EvSend(E_ALoc, loc, &eventsource)
```

5.4 Integrating Visualizations Using Subwindows

The Eve program runs each monitor as a standalone application which opens and operates its own window; coordination consists solely of sharing a common event stream and target program execution. Coordinating only the input event stream works, but doesn't scale all that well: if n monitors are launched with generic "open window" commands, the user has to spend substantial time each program run dragging and resizing the monitor windows, just so they do not overlap with each other, and allocate screen space to some semblance of the user's priorities. Eve supports a slight improvement upon this anarchy: a cooperative coordination scheme in which each monitor may take command line arguments that tell it what size and position at which it should open its window.

Stronger display coordination is possible if the monitor coordinator creates the subwindows for each monitor on a shared 2D main window. The monitor coordinator creates the main window with an ordinary call to `open()` with mode `"g"`. For each monitor, it then creates a subwindow using `Clone()` mode `"g"` on the main window for each monitor, and assigns the monitors' `&window` variable using the subwindow returned by `Clone()`. In the monitor coordinator:

```
w := open("Dashboard Eve","g", "size=1280,1024") | stop("can't open window")
every m := !Monitors do {
  m.subwin := Clone(w, "g", "pos="||m.pos, "size="||m.size)
  variable("&window", m.task) := m.subwin
  }
```

Of course, someone has to specify the monitors' positions and sizes. They might be read in from a configuration file, which was generated by a drawing/layout tool at some earlier time, or they might be algorithmically derived by a function that takes in the monitors' number, priority, and size and aspect ratio constraints.

Although integrating monitors into a unified screen using multiple 2D views in subwindows is an interesting and worthwhile endeavor, and might be the type of presentation you'd expect to see if incorporating visualizations into an integrated development environment, the focus of this book is on incorporating those views within 3D scenes, which is the subject of a later chapter.

References

1. S. P. Reiss, "Connecting Tools Using Message Passing in the FIELD Environment," *IEEE Software*, pp. 57–66, #jul# 1990.
2. D. Garlan and E. Ilias, "Low-cost, Adaptable Tool Integration Policies for Integrated Environments," in *Proceedings of the Fourth ACM SIGSOFT Symposium on Software Development Environments*, 1990.

Chapter 6
Sharing Visualizations Across a Network

This chapter addresses the network communications requirements of visualizations that are to be depicted on multiple devices. While a general primer on data networks is beyond the scope of this text, the network facilities in Unicon are easy to use, and most programmers will find the material accessible even if they have not had a previous background in networks. The goal of this chapter is to present simple techniques that would allow a programmer with a visualization tool such as those found in Chap. 4 to provide the graphical output identically on two or more machines. The communications described here may be considered a necessary prerequisite for the ability to display such visualizations within a networked virtual environment as described in the following chapters of this book.

6.1 Connections: Firewalls, Hole Punching, and Tunneling

The modern internet has two primary communications modes, the connection-based TCP byte stream protocol and the connectionless UDP datagram protocol. There are a variety of circumstances under which UDP is advantageous, some of which arise naturally within virtual environments. However, many or most applications use TCP for some of their communication needs, and it is somewhat simpler to learn and use. We will consider TCP primarily in this book, leaving UDP as a subject for future study.

TCP connections are established between a client that initiates the connection, and a server, which listens for and accepts such connections. Once a connection is established it is bidirectional and the roles of client and server are completely open for the application to define. To request a connection, a Unicon program calls `open(destination, "n")`, where `destination` is a string in `machine:port` format and the machine prefix is either a domain name or an IP

C. Jeffery, J. Al-Gharaibeh, *Writing Virtual Environments for Software Visualization*, DOI 10.1007/978-1-4614-1755-2_6

number. At the receiving side, Unicon's `open()` function has a TCP *networking accept* mode (`"na"`) that waits for incoming connections, but it is usually more practical to use the non-blocking network *listener* mode, indicated by a call to `open(source, "nl")`. Such a listener is generally passed along with zero or more active connections into a function called `select()` that allows a server to handle I/O on any active or new connections at the same time. A listener is automatically converted to a normal TCP connection when an incoming client requests a connection.

6.1.1 Firewalls

TCP connections are fully symmetric and allow either client-server or peer-to-peer communications, but in order for any communication to take place, the two machines have to know each others' IP addresses, and all the network devices in between must be willing to let the data traffic through. One of the tricky parts of modern internet communications is that anyone operating a public site that allows incoming connections gets attacked almost continuously by malicious software. Consequently almost all machines other than those of full-time professional service providers tend to be inaccessible behind firewalls. If you administrate your own network router to access the internet, you may configure it to allow incoming connections on specific ports. The first 1024 ports are generally used for standard operating system services, but applications such as visualization may be free to use a higher port (such as 2000 or 4000) if the firewall will let traffic through.

6.1.2 Resorting to Existing Protocols

If you can't open up traffic on your router, you may be able to develop a visualization tool that works over the network using an existing standard protocol that your router might happen to support, such as the port for a HTTP service (port 80) or for a MySQL database (port 3306). A more extreme example of this is SSH Tunneling, described in its own section below.

6.1.3 Hole Punching

The vast majority of machines do not have a public IP address; instead they have a local IP address assigned by a local DHCP server and reach the public internet via a network address translation step. In a typical ostensibly peer-to-peer scenario, if user A wants to show the output of a visualization tool to user B and they are not on the same local network, they face quite a conundrum. Typically a public directory server of some kind will be involved in helping the two machines find each other, even if the resulting communication is peer to peer. And for peer-to-peer communications, a hole punching technique will be involved in establishing the connection between the peers in spite of firewalls that hinder such communications. The typical sequence looks like:

- Clients A and B contact a directory server, which tells each client the others' external/public IP address.
- Clients A and B try to initiate connections to each other.
- One or the other of the two firewalls generally allows the incoming traffic, since they saw their own machine send out a message and believe they initiated the communications

6.1.4 Tunneling

Many machines that won't allow other types of incoming connections, will allow the SSH secure shell on port 22, since it requires valid username and id, and is encrypted. SSH can be configured to allow other applications' communication to be transmitted using an SSH connection as an encrypted tunnel. The SSH communications are forwarded to a different port, which the application connects to and uses as an ordinary TCP connection—the SSH tunnel is transparent and requires no code internal to the program using it.

6.2 Network Protocols and Message Types

Network protocols define the format, number, and meaning of messages that are transmitted over a network for a particular application. This section describes general TCP-based communications in Unicon. Network protocols resemble file formats, and complex messages may be parsed using techniques similar to those used for reading files.

6.2.1 Transmitting Strings and Structures

TCP applications are often based on human-readable (or semi-readable) text communications. Although they are capable of transmitting arbitrary binary structures, most applications avoid binary representations due to their lack of portability on CPUs with varying word sizes or endianness. The main issue in defining a string-oriented message format is to define message sizes or delimiters. Some protocols use newlines, while others might transmit a size at the beginning of each message.

6.2.2 Serialization

For simple structures (say, a list of integers for example), a simple string message format is adequate: perhaps the list might be a string containing values surrounded by square brackets and separated by commas, for example. For more complex structures, a serialization may be needed. Serialization is an encoding scheme in which a structure is converted into a string in a fashion that preserves pointer relationships and the contents of substructures. Generally, each substructure is represented once and labeled with a unique identifier. Pointers to substructures are converted into identification labels. In the extreme case of a structure that contains pointers back to itself, a serialization will handle such cycles. Unicon has a library, xcodes, that serializes arbitrary structures composed of a mixture of lists, tables, records, sets, and scalar types such as strings and numbers.

6.2.3 Buffering

Transmitting data is fairly easy in applications software, but network hardware and operating systems' network software stack can only transmit a limited number of packets per second. This number is often surprisingly low, especially when packets involve synchronous round-trip communications between applications on two different machines. You might think your hardware should support a million packets per second, or at least ten thousand packets in a second anyhow, and you might find out that in real life the number is way lower. Observed throughput might be a few dozens or hundreds of applications packets per second passing between server and client.

In order to handle many different kinds of visualizations, or many different users (in a virtual environment) it may be essential in the application to bundle several or many messages in each packet. For example, you may decide to send packets 60 or 120 times per second, with however many program events there are to visualize in that amount of time, instead of sending each one in a separate packet. Monitors will not update more often than that anyhow. The decision to bundle multiple messages per packet is generally synonymous with the decision to create a buffer in which messages are stored up until it is time to transmit a packet.

6.3 Messages that Convey Graphics Output

For basic graphics output, one might construct a string corresponding to each graphics function call. If a window parameter is included, a label is needed to identify the corresponding window on the receiving machine. For example a call to `DrawLine(w, 10, 10, 100, 100)` might be converted to "`DrawLine(w_1, 10, 10, 100, 100)`" and transmitted over the network with either a size prefix or a newline to indicate the end of the message. For this book, we will use a protocol that is mostly human readable text. The protocol is line-oriented, so a newline indicates the end of a message. If a newline were part of a message, we would have to escape it (e.g. use \n in the network message). Furthermore, we will append a common prefix to all messages: "\rpc ". The complete network message to draw a line thus becomes:

```
\rpc DrawLine(w_1, 10, 10, 100, 100)
```

The prefix, which starts with a backslash in homage to classic MUDs and MMOs such as EverQuest that begin network messages with a slash or a backslash, provides a measure of verification that one is really looking at the start of a message, and allow these visualization-oriented graphics output messages to be embedded within a larger protocol for a collaborative virtual environment. The "rpc" part of it stands for remote procedure call; the graphics output messages are a simplified special case of a more general problem of executing code on a remote machine.

Unicon's 2D facilities have a grand total of 48 functions, and its 3D facilities add 23 more, for a total of 71 functions. Effectively there are 71 submessages within the "`\rpc`" message, representing these 71 built-in functions. The tricky parts about this whole protocol are the window identification numbers: the w_1 in the example. Each window opened in a Unicon program has a unique serial number sufficient to identify that window in that program run, but those serial numbers usually refer to different windows on other programs running on different machines. Every shared window needs a global identifier, separate from the serial number. Such global identifiers must be created in varying circumstances under different applications. They may be statically assigned in advance, or created on the fly as needed.

The entire contents of a 3D scene might be transmitted by taking a serialization of its display list, a list whose elements are lists and records that encode all output to a window. Transmitting the entire scene may be the optimal way to allow new clients to connect to an existing session. However, retransmitting the entire scene over a network on every frame does not scale well.

6.4 Messages for Events and Program Behavior

There are two kinds of events that need to be encoded and transmitted over the network. User interface events, and program execution behavior events. User interface events, complete with timing and meta-key information, are generally encoded by Unicon as three Unicon values on a list. If these three values are serialized, they can be transmitted together and interpreted on a remote display.

Program execution behavior events pose a more significant challenge. Each event is a simple (event code, event value) pair that could be serialized, but event values can be arbitrarily large, since they can denote arbitrary program values. For local monitors this poses no problems because the monitor runs inside the same address space and can access the structure natively, but for a remote monitor it is a big problem.

Networked visualizations will fall into four categories: (1) the events requested by the monitor have only small scalar event values that can be transmitted as-is, (2) the visualization requires access to a definable subset of the events' structure values, such that the structures can be encoded and transmitted practically, (3) the remote monitor models the structures locally, so that with a mapping from structure id to model id, the network messages need not transmit entire structures, only their ids, or (4) the visualization requires entire structures to be transmitted, and users are willing to wait for results, or willing to live with scalability limits.

6.5 A Remote Memory Allocation Visualizer

This section presents a distributed version of the memory allocation strip chart from Sect. 4.4.3. The range of events to be transmitted are `AllocMask ++ E_Collect`: 18 events that have integer event values. With this restricted event set, network transmission of the events and especially their event values pose no great problem. Effectively, the simple monitor that processes program events and graphs them at the same time is split into two programs (herein referred to as the monitor and the visualizer), connected by a network. The monitor is almost unchanged except that it sends network messages instead of making graphics calls. The function to transmit an event over the network from the monitor to the visualizer is

```
procedure transmit(n, eventcode, eventvalue)
  write(n, eventcode || eventvalue)
end
```

The visualizer is an extremely simple program whose main loop reads from the network instead of calling `EvGet()`. The procedure to unpack an event in the visualizer is

```
procedure getmsg(n)
  msg := read(n) | fail
  msg ? eventcode := move(1) & eventvalue := integer(tab(0))
end
```

Part II
Virtual Environments

Chapter 7
An Overview of Virtual Environments

Different people define the term "virtual environment" differently. For our purposes, a virtual environment is a 3D space within which users can share experiences and information. As the technological barriers to creating virtual worlds have decreased, researchers have created many collaborative virtual environments to serve various domains. The popularity of virtual worlds has increased, as well as their numbers and users, with hundreds of virtual worlds available online populated with tens of millions of players. In this book the types of experiences and information that are of interest are generally graphical in nature: visualizations of programs and program executions. This chapter provides a general introduction to multi-user 3D virtual environments suitable for this domain.

- Features and Usage Patterns
- CVE: a Brief History
- Users and Avatars
- 3D versus 2D Collaborative Views
- Communications Primitives
- Virtual Objects

7.1 Categories

The term virtual environment is used to refer to a wide range of applications. These applications can be categorized using different criteria as explained in the following:

- Based on the number of users:
  - Single user: The popular game Super Mario Bros is an example [1]
  - Multi-user: a limited number of users (in most cases less than 20) can share the virtual world over a local network. An example from games includes Doom [2].
  - Massive multi-user: a large number of users (thousands or more) share a huge virtual world connecting over the internet, such as Second Life [3] and World

C. Jeffery, J. Al-Gharaibeh, *Writing Virtual Environments for Software Visualization*, DOI 10.1007/978-1-4614-1755-2_7

of Warcraft [4]. This is distinguished from Multi-user above by the large number of users, and also the persistent world which continues to exist and change while the players are not online. The world itself continues to be online.

- Based on appearance:
  - **Textual**: the world is described by words leaving the interpretation to the user's imagination. MUDs (Muti-user Dungeons) are examples of such worlds. AberMUD is an example MUD game.
  - **2D**: the world is represented by 2D graphics. Examples include the old arcade games and also many Adobe Flash games.
  - **3D**: the world is constructed via 3D graphics. Many popular virtual worlds fall in this category, such as Second Life [3], ActiveWorlds [5] and many other games.
- Based on technology:
  - Desktop virtual environment: the virtual world is a conventional application, no special hardware is used. Most computer games belong to this category.
  - Virtual reality: includes enhanced elements of immersion, assisted by hardware and devices, such as gloves and head mounted displays. Usually this is used for specific domains like manufacturing and training such as flight simulators (Fig. 7.1).
  - Augmented reality: Augmented Virtuality and Mixed Reality are combinations between real and virtual environments. Mixed Reality refers to any environment that has both real and virtual aspects. If this environment is real with added virtual aspects of information it is called augmented reality. If it is virtual with added real aspects it is called augmented virtuality. For example, augmented reality starts with a real physical environment and the computers alter the view of this environment by adding or removing information (visual, sound, etc…) on top of this view. A hypothetical example would be the ability to walk downtown wearing glasses that block all ads replacing them with

Fig. 7.1 Virtual reality: flight simulator. (source: www.freeway.org)

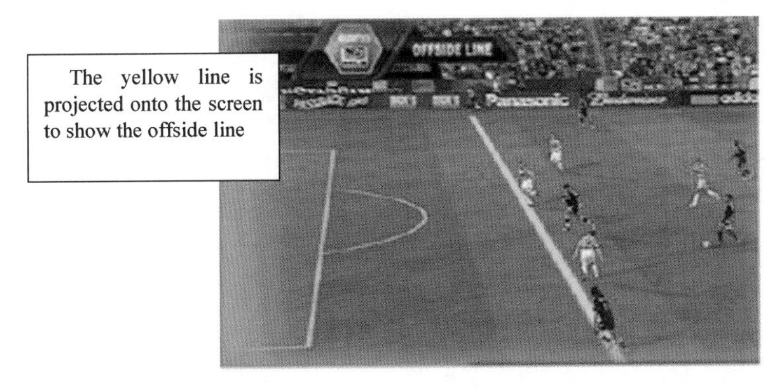

Fig. 7.2 The use of augmented reality in a sport show. (source: www.totalfootballmadness.com)

a photo of a tree. Augmented reality is widely used in military, industry and scientific research. It is also used in sports, while watching a soccer match for example; TV viewers can see lines, numbers and so on projected onto the soccer field explaining an "offside" situation (Fig. 7.2) or how far a free kick is from the target.

7.2 Features and Usage Patterns

The genre that has popularized virtual environments is the MMO—the Massively Multiplayer Online game. This genre as exemplified by EverQuest, begins with the following features:

- the virtual environments we are interested in are at least as immersive and 3D as Doom, the first mass-market immersive 3D game whose graphics were good enough on regular hardware to make people feel presence, motion sickness, etc.
- virtual environments that are of interest for this book are ones that support a large number of users over the internet. Although interesting applications may be constructed for small communities, the ideal virtual environments are relatively open and scalable.
- any virtual environment worth its salt enables users to manipulate dynamic state, and to see each others' manipulations of state, either at the time of occurrence or subsequent login.

7.3 Users and Avatars

Virtual environments depict other users' online presence and activities, generally by means of humanoid avatars. Some virtual environments emphasize avatars and their appearance much more than others.

Beyond avatars and depictions of current activity, virtual environments typically incorporate elaborate user models of experience and skill, derived primarily from logs and analysis of users' actions within the virtual environment. Games tend to model virtual experience gained doing virtual activities, while educational virtual environments may model users' demonstrated knowledge and skills on real-world subjects and tasks, to the extent that those can be demonstrated within the environment.

7.4 Collaborative Views

A virtual environment provides a collaborative view of a shared 3D space to those users who are logged in to the same virtual space at the same time. Users at different locations may see different virtual objects, or a shared object at different levels of detail, from the different points of view afforded by their position and the direction they are facing.

In addition to shared views of static and dynamic 3D objects, the textures of objects rendered all constitute 2D views. In most virtual environments this might include text or 2D signage. In this book such 2D textured rectangular objects are also used in order to depict 2D visualizations of program behavior.

7.5 Virtual Objects

A virtual object is an entity in a virtual environment that has the property that a user can in some ways and under some circumstances interact with it or operate it in some way. This definition distinguishes a virtual object from a purely decorative graphical entity that the user might see but not touch or use. Many software visualizations may be non-interactive graphical renderings, but more powerful ones will introduce virtual objects into the environment, allowing the user to obtain more information, or connecting them with the data being depicted.

7.6 CVE: A Brief History

The virtual environment used in this book as a host within which to embed software visualizations is CVE, the Collaborative Virtual Environment hosted on SourceForge at `cve.sf.net`. CVE was begun around 2004 at the New Mexico State University in support of a project to provide distance education for Computer Science courses. As such, it provided a simple 3D environment within which users could interact via text and voice chat, as well as via a collaborative integrated development environment.

A primary side-effect of the development of CVE from 2005–2009 was the improvement of the development language, Unicon, in support of the application domain of virtual environments. The networking facilities were greatly enhanced to scale to larger number of users. The graphics facilities were refined to support lower-end graphics devices, and greatly extended with transparency, dynamic textures, 3D object selection, and 3D font support. In 2007 work on CVE moved to the University of Idaho. Its scope expanded in response to user feedback: besides collaborating on computer programs, users needed more things to do in the virtual environment. A quest system and non-player character architecture were developed. By 2012, CVE's developers began to look at additional application areas besides Computer Science education. Collaborative Software Visualization is one such application. Although at present CVE's visualization capability consists of embedded visualizations within 3D virtual environments, it may well develop into a more fully-integrated concept in which monitors visualize program execution behavior directly as 3D objects within the virtual environment.

References

1. "Nintendo's Official Home for Mario," Nintendo, [Online]. Available: http://mario.nintendo.com/. [Accessed August 2012].
2. "Doom (video game)," [Online]. Available: http://en.wikipedia.org/wiki/Doom_%28video_game%29. [Accessed August 2012].
3. "Second Life," Linden Research, Inc, [Online]. Available: http://secondlife.com/. [Accessed August 2012].
4. "World of Warcraft," Blizzard, [Online]. Available: www.worldofwarcraft.com. [Accessed August 2012].
5. "ActiveWorlds," ActiveWorlds Inc, [Online]. Available: http://www.activeworlds.com/. [Accessed August 2012].

Chapter 8
Virtual Worlds Graphics and Modeling

In this chapter:

- Static Elements: file format and graph structure
- Adapting Textures from Photos and Drawings
- Dynamic elements: updates
- Living Cities Metaphor: generating a world model from a code repository

8.1 3D Graphics and Modeling 101

The huge advances in graphics hardware in the past two decades allowed for a dramatic leap in the degree of realism that computer graphics can achieve, with richer scenes and larger virtual worlds. However, this also increases the amount of data to manage and code to organize, exploding the complexity of such applications. Computer graphics sits at the center stage when it comes to games and virtual environments. Graphics programming is probably the hardest task in developing such applications both in terms of programming complexity and performance requirements.

Modeling means different things to different people. In some contexts a model is any entity that usably or recognizably approximates another entity. To many computer folks, a 3D model would be a set of textured polygons, optionally with some predefined or hardwired animations of selected portions, stored in a data file, and produced with a tool such as 3D Studio Max or Blender. In this book we are interested this latter sort of model, but we take a broader approach for two reasons. First, many specialized modeling tools are expensive and have a steep learning curve that makes them unattractive for our purposes. Most of the popular file formats are proprietary. Second, modeling tools embody a high quality but labor intensive manual content creation approach that is the opposite of what we need in order to depict large amounts of dynamic data generated on the fly under program control. For our purposes, the virtual environment's 3D model is a dynamic, memory-based data structure, constructed by reading any number of static and dynamic data sources.

C. Jeffery, J. Al-Gharaibeh, *Writing Virtual Environments for Software Visualization*, DOI 10.1007/978-1-4614-1755-2_8

8.2 The Programming Language as a Graphics Engine

Unicon, like many other very high level languages such as Python and Ruby, provides very powerful data structures like lists, tables and sets that are very suitable for representing the virtual world state and data. Unicon also provides high-level APIs for graphics and networking facilities [1, 2]. These features make it an attractive language for writing virtual worlds. Although language extension is desirable in many cases, each addition to the language required for virtual environment features should be small and have a minimum impact on the language and its performance, especially if the change is visible at the language level, such as adding a new function. Different parts of the language have been extended or improved throughout the years to meet new requirements, namely the requirements of the CVE virtual world. Some of these improvements and additions are still ongoing. The next several sections highlight some of these major extensions.

8.2.1 3D Graphics API

Unicon's 3D graphics facilities take many burdens off the programmer. Unlike OpenGL, Unicon has a built-in support for mainstream window systems. Opening windows and handling input events comes free of any extra work.

Despite the high level of the original Unicon 3D API, several features were either missing, or did not meet the performance requirement while developing the CVE virtual environment. These features were improved and extended over time to include more capabilities. This includes a way to manipulate the OpenGL matrix stack, support for dynamic texturing, texture buffering/caching for better performance, JPEG and PNG image file format support for textures in addition to the language supported GIF format, vertex normals support for better smooth shading, and several other improvements including 3D selection.

Unicon inherited Icon's 2D graphics [3]. In addition to several enhancements Unicon added to 2D graphics, it also added support for 3D graphics in 2003 using OpenGL as the underlying library [4]. The 3D facilities in Unicon provide a high level subset of OpenGL's capabilities including drawing primitives, transformations, lighting and texturing.

Many of the new 3D graphics features are extensions of Unicon's 2D API. Unicon reduces the number of functions needed compared with the standard OpenGL C interface. Instead of the more than 250 OpenGL [5] functions that C programmers have to learn, Unicon provides about 30 3D functions. For example to draw a point in OpenGL, a call to `glVertex()` must be made between `glBegin()` and `glEnd()`. Depending on the argument number and type, `glVertex()` takes several forms such as:

```
glVertex2i()    glVertex3i()    glVertex2iv()    glVertex3iv()
glVertex2f()    glVertex3f()    glVertex2fv()    glVertex3fv()
glVertex2d()    glVertex3d()    glVertex2dv()    glVertex3dv()
```

These functions are replaced by one function in Unicon, `DrawPoint()`. The Unicon runtime system handles the different data formats. This is the case for many other functions in OpenGL.

Unlike OpenGL, where the program has to have a `display()` function to keep track of all the scene content and draw it again every time the screen needs to be refreshed, Unicon does this job implicitly. For this purpose, Unicon attaches a display list to each opened 3D window (Fig. 8.1). This list is just a regular Unicon list containing all of the information needed to recreate the whole scene. Every time a new primitive is added to the scene, color is changed, or a transformation function is called, a corresponding element is added to the list remembering what needs to be done, where, and any other information needed with that element.

The display list becomes large for complex scenes, limiting the scalability of 3D programs. Another function named `WSection()` provides a simple form of scene partitioning. The function allows the programmers to create sub slices that can be marked to be skipped during rendering. When developing a virtual world, this feature enables things like skipping rooms that are either far from or not visible to users.

`WSection()` expands the scalability of Unicon programs but does not solve the 3D performance problem for scenes with very high polygon count, especially those with 3D models loaded from model files. One limit to performance in such scenarios is the way data is represented in the display list. For a large mesh, thousands of vertices need to be stored in the display list element that corresponds to that mesh. This element is also a Unicon list. Every time the scene needs to be refreshed the data has to be converted to the appropriate data format and copied to a C array. This array is then passed down to OpenGL for final rendering. For scenes with complex meshes, this repeated conversion and data copying takes most of the execution time, consuming more than 80 % of the time in some experiments. This was a great opportunity for improving the performance of programs of this kind, which applies to many graphics applications including virtual worlds.

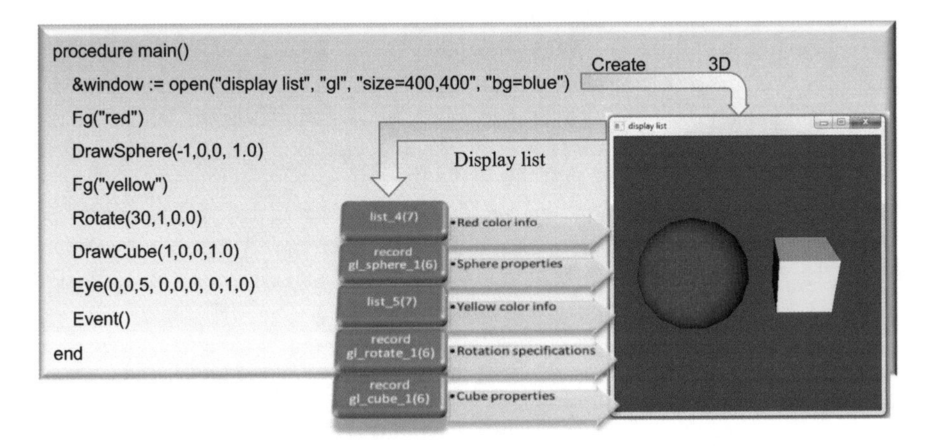

Fig. 8.1 A 3D window in Unicon keeps track of all the scene content using a display list

8.2.2 Improving 3D graphics performance

The 3D performance of programs written in Unicon represents a compromise between the underlying C OpenGL code of the virtual machine runtime system, and the flexibility and ease of programming afforded in the higher-level language. Performance can be lost due to dynamic language representations of data, or by the language's hard-wiring various parameters of the OpenGL semantics.

8.2.2.1 Data representation

3D graphics makes extensive use of integer and double data types. A 3D model for example, might contain tens of thousands of double and integer numbers for vertex data, indices, texture coordinates, animation and more. These kinds of data are usually stored in arrays that get passed to OpenGL for final processing and rendering. Unicon's list data type is ideal for storage and manipulation of such data at the language level. Unicon lists are not arrays; they are more general and more powerful. A list can store heterogeneous data types, and can grow and shrink. This means that lists have a different representation and implementation than that of the C arrays used by OpenGL. While their representation makes lists very flexible and easy to use, it also means the data stored in a list cannot simply be passed to OpenGL; it has to be converted to an array format first. The underlying implementation of 3D graphics in Unicon was improved to avoid repetitive conversion for the same data from one frame to the next if the data does not change. The conversion is still necessary whenever the data changes for a specific object, which is the case for many animated objects.

8.2.2.2 Arrays as Lists

Improving performance by buffering or caching the data converted from a list to an array is only a partial solution. Buffering imposes a memory overhead that might be large for rich scenes; also it does not work for any dynamic object in the scene that requires frequent update to its data. A more general solution is needed that does not require more memory and works well for any objects, including those that involve animation. Animation makes a difference because usually animation is done by key framing and applying an animation transformation to vertex data, generating a new set of world vertices that replaces the old set. In other words, the solution should make the same data visible to both the language level and the underlying OpenGL function, bridging the gap between the language interface and the graphics library.

One way to accomplish this is to add new data types to the language to hold arrays of integer and double data. To have the least impact on the language interface, another route was taken. The list data type was extended to support arrays of data by changing its implementation in the language runtime system. This keeps the design in line with the language spirit and respects a major goal, which is not to have any visible additions to the language interface unless it cannot be avoided.

In the new design of lists, for integer and double data types, an array is just another list that happens to have a fixed initial size and also one type of data, either integer or double. Because of the extensive use of lists in Unicon programs, and to avoid any unintended side effects when using arrays, a temporary new constructor function for arrays is added. A "regular" list is created using the `list(size, initial value)` function. An "array" style list can be created using the new `array(size, initial value)`. The initial value data type (integer or double) dictates the data type of the returned array list. Beyond the creation of a list, whether "regular" or "array", all other operations are the same. The following code fragment creates two lists of size 10 and initializes their elements from 1.0 to 10.0.

```
L := list(10, 0.0)
A := array(10, 0.0)
every i:=1 to 10 do  { A[i] := L[i] := real(i) }
```

Any non-array operation that is applied to an array list forces it to be converted to a regular list. For example applying `pop(), push(), get()` or `put()` at the array `A` in the code above. This is done by the runtime system without programmer intervention.

References

1. C. Jeffery, S. Mohamed, R. Parlett and R. Pereda, Programming with Unicon, 2003.
2. R. Griswold and M. Griswold, The Icon Programming Language, 3rd ed, San Jose, CA: Peer-to-Peer Communications, 1999.
3. R. Griswold, C. Jeffery and G. Townsend, Graphics Programming in Icon, San Jose, CA: Peer to Peer Communications, 1998.
4. C. Jeffery and N. Martinez, "The Implementation of Graphics in Unicon Version 11 (Unicon Technical Report #5a)," 2003.
5. D. Shreiner, M. Woo and J. Neider, OpenGL(R) Programming Guide: The Official Guide to Learning OpenGL, Version 1. 2, 3rd ed, Amsterdam: Addison-Wesley Longman.

Chapter 9
Non-Player Characters and Quests

In this chapter:

- Computer-controlled characters and their roles in MMOs
- Knowledge, dialog, and behavior models
- Quests as a way of sharing knowledge
- Quests as a way of asking for collaborative participation and assistance

Non-player characters (NPCs) are an important component of role playing games including massively multi-player online games (MMOs), presenting many of the activities and much of the storyline of the game. Computer-controlled NPCs can be enemies, allies, monsters, or pets, but their most important role is that of quest giver or assistant to the user who goes on an adventure. In virtual environments for software visualization, the primary goal of the user is often program understanding or debugging. Non-player characters might embody certain potential bugs or hypotheses about program behavior that the user might want to explore, or they might be there to assist the user in interpreting the visualizations being shown.

CVE provides a simple NPC architecture called PNQ (Portable Extensible Non-player characters and Quests). PNQ is a prototype; only essential parts are implemented, which make such NPCs viable test subjects for the purpose of this book. These essential parts include basic and partial support for knowledge, dialog and behavior models described in the following sections. The NPCs can keep track of the quests they have, the quests each user accomplished, can carry on very short conversations with predefined chat messages, and have very basic wandering capability. Providing a full and robust NPC implementation is outside the scope of this book.

The main purposes of these NPCs are to add actions in the CVE virtual world, stress both the clients and the server and test new language features. NPCs in CVE not only enrich the virtual world, they also open the door for more use cases and test scenarios once their implementation is complete. PNQ architecture is independent from the CVE architecture, and can be run as independent processes connecting to the server as regular clients. PNQ NPCs can also be run inside the virtual world

C. Jeffery, J. Al-Gharaibeh, *Writing Virtual Environments for Software Visualization,* DOI 10.1007/978-1-4614-1755-2_9

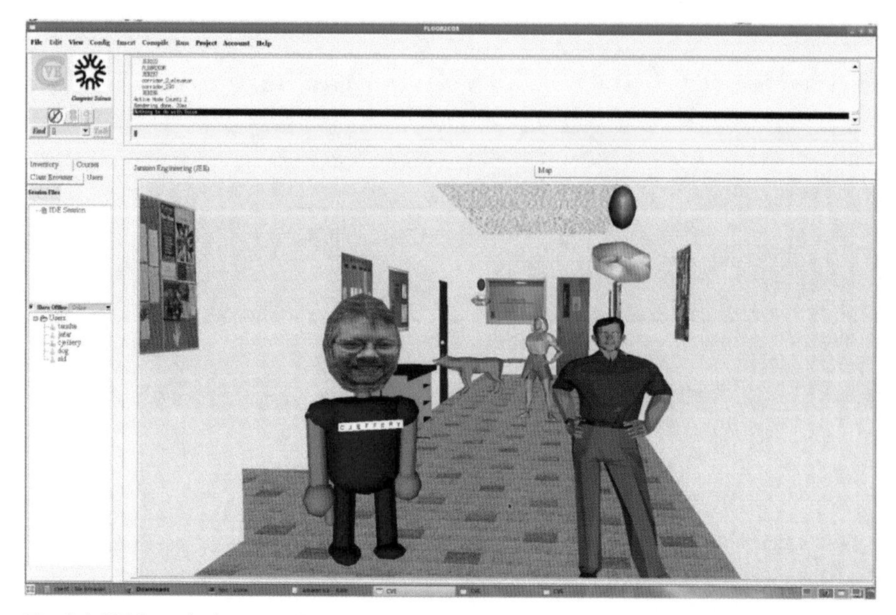

Fig. 9.1 NPCs and other users in CVE. An NPC is marked with red ball above their heads

server if the server supports such integration as is the case for the CVE server presented later in this chapter. Figure 9.1 shows several avatars from the CVE virtual world. A quest giver NPC in CVE has a red ball over his head. This marks NPCs with available quests as persons in the environment with whom the user has special reason to interact.

The addition of NPCs triggered many new language features added to Unicon. In most aspects, NPCs are similar to regular users, putting more pressure on the virtual world server and the client. For CVE that means finding new ways to improve scalability in terms of the number of users it can handle. Similar things can be said about the client, especially that NPCs brought with them 3D models which put extensive pressure on the Unicon 3D graphics rendering pipeline, requiring new features such as array and thread support.

This chapter presents the PNQ architecture including NPCs and their quest system along with a summary of their design and prototype implementation. The chapter briefly introduces NPCs in the CVE with 3D models support in the CVE client followed by the integration of NPCs in the CVE server.

9.1 NPCs

A PNQ NPC is created much as a regular (human-controlled) character. End users can utilize their regular account or an auxiliary account to create an NPC character, adding quests and activities and making them available to other users. The intention is that when the user is logged off, their avatar is still present on the system,

Table 9.1 The major parts in an NPC profile file

Section	Description
Id	An "ID card" presentation of the NPC, suitable for an "inspect details" operation in a game. The id provides an image, name, and basic attributes
Knowledge	A specification of the NPC's knowledge model consists of a teaching section with a bulleted list of named links to quests. NPC knowledge also includes a more dynamic experience (user model) database that is not part of the profile
Dialogue	A specification of this NPC's verbal capability, and thereby its personality
Behavior	A specification of this NPC's active (e.g. mobile) behavior. The four kinds of PNQ behavior specification include stationary, wanderer, routine, and companion
Avatar	A specification of this NPC's avatar (link to 3d model file, dimensions, and textures)

controlled from a remote NPC client just like a regular user, interacting with other users as instructed by the player.

9.1.1 NPC Profiles

An NPC profile is a file containing NPC details in simple HTML format. An NPC profile can be created and maintained as a webpage. In HTML they are each given in a named anchor tag. Although a graphical wizard for creating profiles is available, many NPCs can be created manually by copying a template and changing the content details. Table 9.1 lists the major parts in an NPC profile along with a short description about each part.

9.1.2 Knowledge Model

PNQ NPCs use two types of knowledge: the quests they offer, and what they remember about other users from past quests. The former is almost static, refined occasionally by the NPC creator. The latter is dynamic and accrues during game play.

9.1.3 Dialogue Model

In World of Warcraft, players interact with NPCs using popup menus. This contrasts egregiously with human player interactions. Some MMOs such as Everquest enhance the realism of NPCs by forcing NPC interaction through the chat channel. NPCs that chat feel more natural but are frustrating when they fail to respond usefully to a player conversation. Chat-based interaction is more portable across virtual worlds. The PNQ NPC dialogue model consists of chat rules written in AIML [1] augmented with offers to undertake available quests. The AIML rules determine how the NPC replies to player utterances

9.1.4 Behavior Model

The PNQ NPC behavior model provides rules for NPC movements and responses to external stimuli. Four NPC behavior types are supported: stationary, routine, wanderer, and companion. A stationary NPC does not move from a specified home location. A routine NPC regularly does specific tasks including movements at predetermined times. A wanderer is an NPC that moves randomly within a prescribed domain. A companion NPC accompanies a player on a destination-based quest. A behavior language is needed to write the rules that describe how an NPC acts in the virtual world. As discussed, ABL is an example of such a language. ABL and other scripting languages were not chosen for this project because they were viewed as too low-level. Many NPC authors may have little or no programming background. Thus, a new high level declarative behavior model language was invented for PNQ. The new language captures simple behavior scenarios that an NPC should be able to do in an educational virtual world, including movements and offering quests. Figure 9.2 shows a fragment of the grammar.

The following is a stationary NPC example. The NPC logs in at 8 AM in room jeb 228 and offers quests from its profile in the "help" topic until 5 PM.

```
Stationary {
   default : {
     login    8:00 in place(jeb 228)
     offer help
     logout 17:00
     }
   }
```

If no "offer" statements appear in a behavior model, the default behavior is to offer all eligible quests listed in the NPC profile. The following is a more complex example that features a Routine NPC. The NPC logs in to the virtual world and walks to different places, provides office hours and "teaches" by offering quests. In

```
BehaviorModel : Type "{" States "}";
Type          : Wanderer | Stationary | Routine | Companion;
States        : State States | State ;
State         : StateName optSchedule "{" Actions "}" ;
Actions   : Action Actions | /* empty */ ;
Action    : Login | WalkTo | SetState | Offer| Redirect | Logout |
        Teleport;
Login         : LOGIN TIME IN Place ;
WalkTo     : WALKTO Place optAT ;
SetState : SET StateName;
Offer         : OFFER Topics;
```

Fig. 9.2 Part of the grammar for the PNQ NPC behavior model

this model, the NPC spends different parts of the day in different states. The state named MWF is scheduled (by a string notation inspired by the UNIX cron(1) tool) to commence at 10 AM every Monday, Wednesday, and Friday.

```
Routine {
  default : { login in place(jeb 226) }
  MWF "0 10 * * MWF": {
     login in place(jeb 005 corridor)
     walkTo place(jeb 226)
     state(officehours)
     walkTo place(jeb121) at 10:30
     state(teach)
     walkTo place(jeb 226) at 11:30
     logout
     }
  officehours {
     offer debug, question
     }
  teach {
     offer CS210, programming languages, Lisp, Flex, Bison
     redirect debug : Dog
     }
 }
```

9.2 Quests

In games, quest activities are used to teach the game itself as well as to entertain. The NPCs and quests described in this chapter may fill an educational role, besides enabling new form of interaction in the virtal world. Quests are a primary mechanism for tutorial learning. Quest specifications resemble UML use case descriptions [2]. The kinds of steps are limited to those observable by NPCs interacting with users. The main differences between a quest and a use case description are that a quest may contain auxiliary content (such as quizzes and demonstrations) that are used to measure completion of the quest steps, and a quest lists rewards for completion, if any. Quizzes and demonstrations will often need to be external references to pools of questions. The difference between quizzes and demonstrations is that a quiz is delivered and answers interpreted by an NPC agent directly, while a demonstration involves an in-world interaction (in this case, a session with a tutorial UNIX command-line shell) that is monitored by an NPC agent. Evaluation of deeper understanding may require offline human evaluation, or fall outside the realm of what an NPC Tutor can reasonably perform.

9.2.1 *Quest Repository*

PNQ Quests are maintained in the same way as NPC profiles: they are HTML files linked from the knowledge section in the NPC's webpage. The quest webpage includes several sections. A quest builder tool facilitates the process of creating new quests. Figure 9.3 shows an example quest from the domain of computing. Figure 9.4 is a screenshot of a quest being offered in a response to a user clicking a quest red sphere above an NPC.

9.2.2 *Quest Rating and User Reward via Peer Review*

Players need to be motivated to perform quests, and their accomplishments need to be recognized. The main kind of reward that matters to PNQ is the experience points in specific topic areas that enable a character to undertake more advanced quest activities. In the ls activity presented in Fig. 9.3, two specific previous quests (tutorials on Files and Directories) had to be completed before the NPC offers the ls quest. Completing the ls quest enables any quest that depends on it specifically, and also awards a point of general UNIX experience.

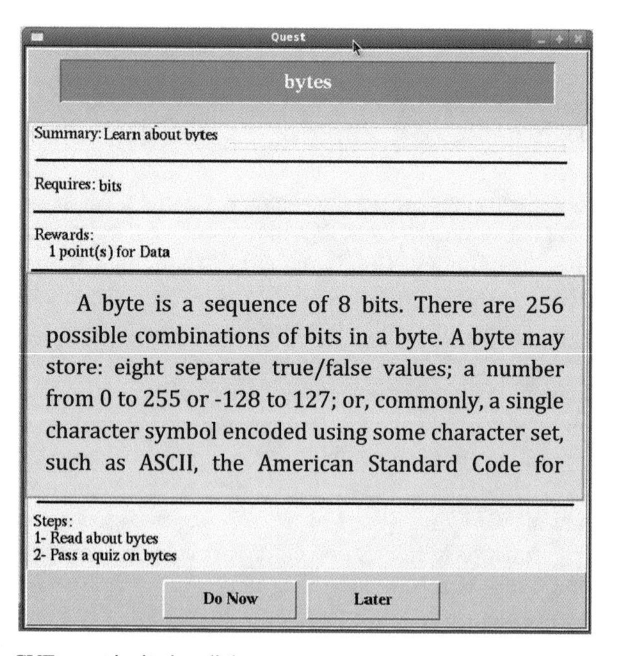

Fig. 9.3 A sample CVE quest invitation dialog

Name : ls
Summary : Learn the basics of the ls command.
Requires : Files, Directories
Steps : Read the UNIX manual page for ls.
Pass a quiz on ls command line options.
Demonstrate "ls" for Tux.
Rewards : UNIX: 1
Quiz (2/2 to pass)
How can you get a long listing that shows file permissions and size?
> ls -l
How can you list all files in all subdirectories?
> ls -R
Demo (2/2 to pass)
Show me a simple listing of the root directory.
> ls /
Show me a listing of the current directory, sorted by the time each file was last accessed?
> ls -t

Fig. 9.4 An example quest as it appears in a web page

9.3 Design and Implementation

The PNQ NPC is designed as a standalone entity; however, they can be plugged in as threads in the server. An NPC client connects to the server like a regular user, with an NPC indicator flag. When run as separate clients, this design frees the server from NPC management, makes the NPC more flexible by allowing the NPC to run from any machine and allows a human to "play" an NPC. As mentioned in the introduction of this chapter, PNQ provides specifications of NPCs and quests and their envisioned design. The implementation details are left for specific implementations.

9.3.1 NPC Architecture

The NPC client design mirrors the NPC profile. The following is the list of the NPC major components: A knowledge engine: composed of two parts, the relatively static quest knowledge and the dynamic knowledge. A behavior engine: dictates how the NPC will behave and move around in the world based on the NPC profile. A chat engine: analyzes the incoming chat messages and generates proper response if possible. Chat messages are categorized either as general chat or messages that involve questions or answers about the quests the NPC is providing. At the top level there is an I/O interface that manages data transfer between the NPC, virtual world server, HTTP servers and also the disk. Figure 9.5 shows the different NPC components.

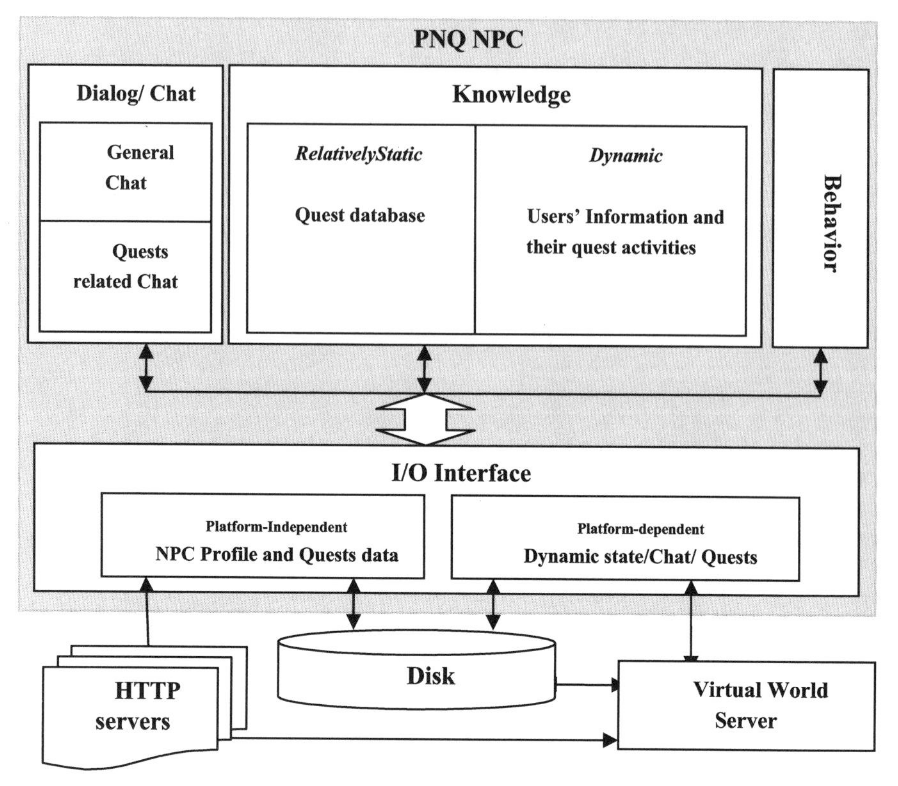

Fig. 9.5 PNQ NPC architecture

9.3.2 Implementation Discussion

The PNQ NPC client is implemented as part of the CVE project. The source code for the PNQ NPC client is available at cve.sourceforge.net. The NPC client is independent from the regular user client in CVE and could be implemented using any programming language. However, Unicon was chosen to implement the NPC client because it provides: (1) a very simple interface to the standard internet protocols such as TCP and HTTP. (2) high level built-in string parsing features that make it easy to parse the NPC profile and quest data. The NPC client starts by downloading the NPC home page and quest pages that define the profile of the NPC. It then initiates a TCP connection with the virtual world server. Upon a successful login, the new NPC is available in the virtual world accepting interactions from other users.

9.3.2.1 Network Protocol

The PNQ NPC network protocol uses string messages consisting of a command name followed by arguments. This enables an easy integration with CVE when

Table 9.2 Summary of the most important NPC protocol messages

Category	Command	Description
User presence commands	users	Brings up a list of all the other users who are currently online
	avatar	Informs about a new user who just logged in
	move	Informs about a specific user movement (x, y, z and direction)
Chat commands	say	Public chat message sent to all users
	tell	Private chat message sent to the intended user only
Quest activities commands	The following commands communicate information about quest activities between the NPC, server and clients. All of these commands are prefixed with (npcmsg quest) and some of them take several arguments containing all of the information necessary for each quest activity	
	LookFor	Asks the NPC for available quests
	Halo	Informs the client that the NPC has available quests
	GiveMe	Asks the NPC to send the next available quest
	Url	Sends the client a quest URL
	Accept	Informs the NPC that the user has accepted the quest
	Cancel	Informs the NPC that the user has cancelled or abandoned the quest
	Done	Informs the NPC that the user has finished the quest

it comes to network communications since CVE also uses string messages. The NPC client recognizes the following messages coming from the virtual world server and other users: chat messages, messages about other users' locations in the virtual world, and messages about quest requests and activities. All commands begin with two forward slashes (//). Some commands have arguments to pass information to/from the NPC. A summary of the quest command and its options is presented in Table 9.2. The different options communicate information about quest activities between the NPC, server and clients.

Figure 9.6 shows a scenario where the quest messages are used to start a new quest. The process starts when a user client clicks the red ball marking a quest over an NPC. The click sends the message "Quest GiveMe" to the server, which forwards it to the specific NPC. The NPC finds the next available quest for that user based on the knowledge it has, then sends back a quest title along with its URL so that it can be downloaded directly from the source webpage. The server gets the message and checks whether the user has already completed the quest or is currently undertaking the quest. The NPC maintains a list of completed and active quests for every user, but if the NPC process gets restarted, the quest protocol allows it to reload user dynamic knowledge on demand. The scenario in Fig. 9.6 shows the case when the NPC needs to get updated. After a quest request message, the server replies to the NPC informing it that this user has completed the quest. The NPC then adds this piece of information to its database and finds another available quest for the user and sends it back to the server, which in turn checks again whether the user has already completed the quest. This time the server approves the new quest and

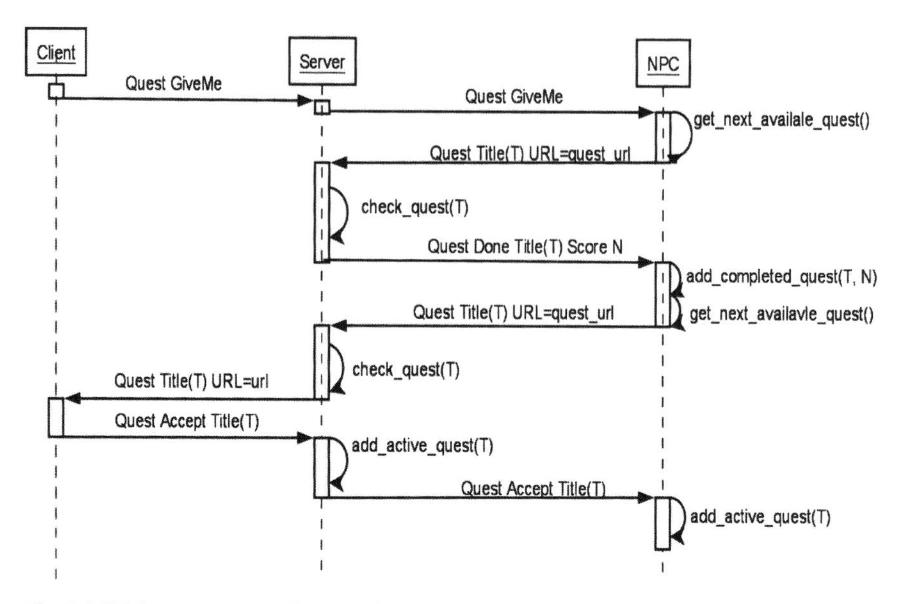

Fig. 9.6 NPC quest messages between the NPC, the server and the client

forwards it to the client. The client gets the message and downloads the quest from the specified URL or loads it from the disk if it is stored on the local disk. If the user accepts the new quest, which is the case in the scenario we have here, it sends the message "Quest Accept Title(T)" back to the server. The server adds the specified quest to the active quest list for the user and forwards the message to the NPC which in turn also adds the quest to the active quests list for that user.

9.3.2.2 Source Code Organization

At the top level view, the NPC source code is organized into two major components: the NPC class and the client application that uses it. The NPC class, called ExternalNPC, features a public interface consisting of `login()` and `mainloop()` methods and a few other methods that control the NPC activities. The NPC class also holds most of the common features shared between different kinds of NPCs such as quest activities and basic chat capabilities. NPC client applications can customize the NPCs and give them distinguishing characteristics beyond what is modeled in the profile web page to any more advanced dialogue capabilities and behavior that are required for that specific NPC. Given the ExternalNPC class, if a new NPC named Tux were created, what does the Tux NPC client look like? Tux has a profile on a web page. Figure 9.7 shows a minimal Tux NPC client to instantiate Tux in the CVE virtual world. Tux's profile is downloaded from the specified homepage. The methods are called to handle the received messages from the server

```
class ClientNPC:ExternalNPC()
method handle_msg(s)
  write("tux received ", s)
end
method idlefunc()
  write("tux is snoozing")
  delay(1000)
end
method run(password)
  srvport := "virtual.cs.uidaho.edu:4500"
  homepage := "www.tux_homepage.com"
  login(password)
  mainloop()
  write("NPC loop finished, good bye")
end
procedure main(arg)
   tux := ClientNPC()
   tux.run(arg[1])  # passing the password
end
```

Fig. 9.7 A very simple and compact NPC client example

and to set what to do when Tux is idle. Although Tux is not doing anything except standing in a fixed position and logging whatever messages he gets from the server without responding to any of them, this is a building block for an NPC client that may form the basis for interesting NPC dialogue, knowledge, and behavior models.

The source code for the NPC client itself is organized into several classes. The following is a list of the major classes along with a summary about each class: ExternalNPC: Holds all the information about the NPC. It has methods for downloading and parsing the NPC profile, managing the connection with the CVE server, receiving server messages and giving proper responses and taking proper actions when asked to chat or to give quest activities. AvatarData: Holds the NPC's knowledge of other users' avatars and their quests activities. The NPC uses this to be aware of other users locations and take proper actions if any. Some NPCs for example might interact with other users if they come closer or go farther, and also avoid them if the NPC is set to move. The NPC uses this also to know what quests each specific user has already completed, quests that they can take and quests that they are currently working on. Knowledge: A collection of knowledge categories. KnowledgeCategory: A collection of quests that belongs to the same category. Quest: Holds all of the information about a quest, such as readings, prerequisites, and steps. It has methods to read and parse a quest from a webpage or a local file and save it if necessary to a local file. The Quest class also keeps track of the quest activities like the current question the user is answering, quest score, and user's answers for different questions and so on.

References

1. R. Wallace, Artificial Intelligence Markup Language (AIML) Version 1.0.1, Working Draft 25 October 2001.
2. G. Booch, J. Rumbaugh and I. Jacobson, The Unified Modeling Language User Guide, Addison-Wesley, 1998.

Chapter 10
Dynamic Texturing in Virtual Environments

This chapter presents:

- Dynamic texturing
- Embedding visualizations of programs.
- Embedding 2D visualizations within virtual "war rooms"
- Embedding 3D visualizations: events as virtual objects

Textures are essential to make realistic scenes. 3D graphics libraries usually have support for texture creation, but once a texture is created nothing much can be done afterward to make any changes to it. In many applications, having textures that can be updated at run time opens the door for new interesting uses. Virtual screens or "live" walls inside a virtual environment can display textures that change over time, simulating a computer screen or displaying visualization directly in the virtual world. In CVE, this was used to implement whiteboards in a classroom. The whiteboard is captured from the instructor side and broadcast to clients, which then update their whiteboards' textures in the virtual classroom.

10.1 Reloading Textures

Dynamic texture support in Unicon does not create a new special case texture that can be used dynamically; rather it adds the ability to use any texture in a dynamic way. Once the texture is changed; all objects in a 3D scene using that texture will reflect the new changes in the texture as soon as the window is refreshed.

One obvious way to update a texture is to reload the texture with a new image. This new image can be a slightly modified version of the original image used to create the texture in the first place. Repeating this process results in an animated texture. This is the same procedure used to create scenes in an animated movie, applied to textures in a 3D scene in this case. The function `Texture(image_filename)` creates a new texture. The function returns a record that holds information about the texture. Another function used to load images, `ReadImage(image_filename)`,

C. Jeffery, J. Al-Gharaibeh, *Writing Virtual Environments for Software Visualization*, DOI 10.1007/978-1-4614-1755-2_10

supports loading new images into a texture. The following code fragment demonstrates the use of these functions:

```
# Create a texture from a jpeg image
Tex := Texture("shot1.jpg")
Refresh()
every i:= 2 to 10 do {
    # wait for a second before updating the texture
    delay(1000)
    # load a new image in the texture "Tex" each iteration
    ReadImage(Tex, "shot" || i || ".jpg")
    Refresh()
    }
```

The function `Texture()` creates as texture `(Tex)` out of the image `shot1.jpg` This code will update the texture `Tex` nine times (`i` loops through 2–10 values) with `shot2.jpg` through `shot10.jpg` images. Of course the above code should be in a context to be meaningful. All of the images should have the exact dimensions if they are meant to replace each other perfectly, but that is not a requirement. Any subsequent image after the one used to create the original texture will be cropped to fit within the original texture size if they happen to be of larger dimensions. Images are all aligned at their origins (0, 0) which is the top left corner of the image. This can be overridden by passing extra parameters to `ReadImage()` in the form of `ReadImage(texture, image, x, y)` where `x` and `y` represents the pixel coordinates in the target texture where the new image top left corner will be placed so that the new image will extend right and down from that pixel.

10.2 Textures as Windows

Using images to update a texture is a useful feature and can be used to implement many functionalities and also produce nice effects. However, the dynamic behavior of textures can be of greater value if it is not limited to the use of images only. One of the uses of CVE requires software visualization to be plotted into a texture in the 3D world. This can be achieved using the technique described in the previous section. An external visualization program can generate the images and feed them to the CVE program, which then can be used to update the texture in the 3D world continuously with the visualization output. While it is doable, it is a slow process requiring a lot of IO operations and involves moving large quantities of data.

A more efficient approach can be utilized if a texture can be thought of as a window. The visualizations in question are 2D graphics and mainly involve simple commands including drawing points, rectangles, lines, and changing colors. If a texture can be used as a window, visualization commands can be written directly to a texture resulting in a very useful, very fast feature. For this new feature, no new functions, and no major changes to the existing 2D graphics API are needed. 2D graphics output are all sent to the current active window unless the first parameter to these functions is a window (or a texture) itself. In that case the output is sent to the window referred to by the parameter.

The dynamic texture feature involves using a texture as a first parameter replacing a window parameter to several 2D graphics functions, this includes:

```
DrawPoint()
DrawLine()
DrawRectangle()
FillRectangle()
CopyArea()
Clone()
```

This represents only the minimal set of functions required to support a software visualization programs in the CVE. These visualizations programs are part of Unicon distributions and make use of the functions above to display their results.

10.3 Dynamic Texture Examples

Dynamic texturing allows the program to update its textures in real time. This can be used to create very nice effects that cannot be achieved otherwise, such as having a virtual computer screen. This section presents examples that demonstrate the use of this feature.

One way to use dynamic textures is to reload a texture with another image. The new image does not have to have the same size or format as the original image. The following program renders a scene with a cube textured from all sides with the same texture. The program pauses for a key press or a mouse click before overwriting the center portion of the texture on the cube with a small image. The result is shown in Fig. 10.1. Once the texture is updated, all sides of the cube reflect the new change.

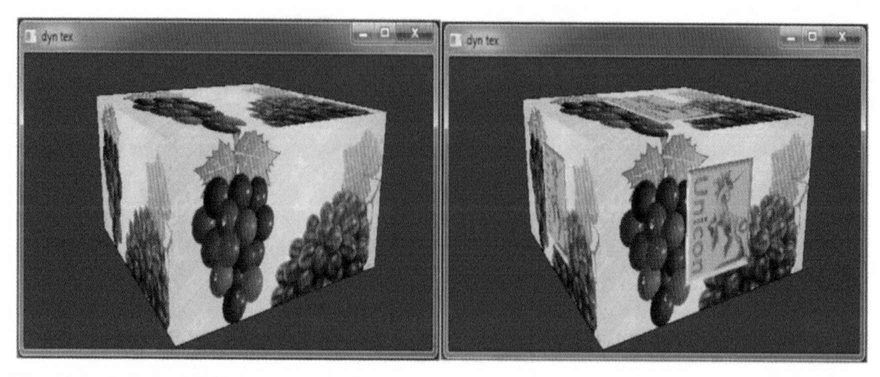

Fig. 10.1 *Left*: the original texture shown on three sides of the cube. *Right*: the texture after getting updated with another image

```
procedure main(argv)
    &window := open("dyn tex", "gl", "size=512,256", "bg=blue)
    WAttrib("texmode=on")
    PushTranslate(0,0,3)
    Rotate(20, 1, 0, 0)
    Rotate(30, 0, 1, 0)
    Scale( 2,1,2)
    tex := Texture("grapes.jpg")
    cube := DrawCube(0,0,0, 2.5)
    Eye(0,-1.0,13, 0,-0.25,0, 0,1,0)
    Event()
    # update the texture with a new image
    ReadImage(tex, "setupbig.png", 174, 46 )
    Refresh()
    Event()
end
```

The new image does not have to be an image stored on disk, it can be captured on the fly from a window. The function `CopyArea()` is extended to support this feature. Instead of copying an area from window to window the function can do the copying directly into a texture in the following manner:

```
CopyArea(win, tex, x, y, w, h, new_x, new_y)
```

In this case, the function copies an area from the window `win` to the texture `tex`. The area copied from win starting at `x, y` coordinates extending `w` width and `h` height. The copied area is placed at `new_x, new_y` coordinate in `tex.` if `new_x` and `new_y` are omitted, their values default to `x` and `y`.

The other use scenario for dynamic textures is to use them as windows, accepting drawing commands directly. The texture in this case is similar to a window's

Fig. 10.2 *Left*: original texture. *Right*: the result of drawing on the texture dynamically

canvas, where lines, rectangles and other 2D primitives can be rendered to be part of the texture. Instead of passing a window as a first argument to these drawing functions, a texture variable is passed, diverting the drawing to the texture itself. The following example demonstrates such use, and Fig. 10.2 presents the screenshots of the program output.

```
procedure main(argv)
   &window := open("dyn tex", "gl", "size=512,256", "bg=blue")
   PushTranslate(0,0,3)
   Rotate(20, 1, 0, 0)
   Rotate(25, 0, 1, 0)
   Scale( 2,1,2)
   WAttrib("texmode=on" )
   tex := Texture("grapes.jpg")
   cube := DrawCube(0,0,0, 2.5)
   Eye(0,-1.0,11, 0,-0.25,0, 0,1,0)
   Event() # pause waiting for an event
   Fg("yellow")
   # Draw 5 rectangles
   FillRectangle(tex,  2, 2, 508, 10,
      2, 245, 508,10,
      2, 2, 10, 250,
      498, 2, 10, 250,
      250, 40, 250 , 70)
   Fg("red")
   every i:=!30 do
     DrawLine(tex, i*8+250,50, i*8+250, 100)
```

```
      every i:=!20 do
         if i%2=0 then{
            Fg("black")
            DrawRectangle(tex,i^2+i*10,i^2, 25,25)
            }
       else{
           Fg("green")
           FillRectangle(tex,i^2+i*10,i^2, 25,25)
           }

      Refresh()
      Event()
   end
```

When using a dynamic texture as window, the difference between the two is blurred at the code level. As shown in the code listing above, the programmers write the code as if the texture is a real window, and the result is reflected on the texture wherever it is used in the scene as illustrated in the figure. This transparency between textures and windows means that programmers do not have to learn a new API, they just apply what they know about windows to textures. It also means that code written to work with windows or textures, requires almost no modifications except changing a variable initialization from a window to a texture or vice versa to have the code work one way or another. A piece of code that needs to support both use cases at the same time need not to be modified or replicated, moving such code to a separate procedure and passing the right parameter is sufficient, further reducing the amount of code needed.

Chapter 11
Embedding Visualizations in a Virtual Environment

This chapter presents the merger of the concept of monitor coordinators from Chap. 5 and network messages for graphics from Chap. 6 into the CVE virtual environment. The major problems to be resolved include: how to identify the dynamic textures being used to embed the visualizations within the virtual world model, and how to put the users in charge of the speed of the visualization.

11.1 Placement of Visualizations in Predefined Locations

CVE's static model is comprised of a set of data files that look like this example:

```
# static properties of a room
Room {
name SH 167
x 29.2
y 0
z 0.2
w 6
h 3.05
l 3.7
floor Rect { texture floor2.gif }
obstacles [
    Box { # window sill
        Rect {coords [29.2,0,.22, 29.2,1,.22, 35.2,1,.22, 35.2,0,.22]}
        Rect {coords [29.2,1,.22, 29.2,1,.2, 35.2,1,.2, 35.2,1,.22]}
        }
```


C. Jeffery, J. Al-Gharaibeh, *Writing Virtual Environments for Software Visualization*, DOI 10.1007/978-1-4614-1755-2_11

```
    ]
  decorations [
    Rect { # window
      texture wall2.gif
      coords [29.2,1,.22, 29.2,3.2,.22, 35.2,3.2,.22, 35.2,1,.22]
      }
    Rect { # whiteboard
      texture whiteboard.gif
      coords [29.3,1,2.5, 29.3,2.5,2.5, 29.3,2.5,.4, 29.3,1,.4]
      }
    ]
  }
```

Within such a model, any virtual object may be used as a visualization site by assigning its texture a global identification. An example is presented in Fig. 11.1. An optional field (id) is used for this purpose. For example, to make the virtual white-

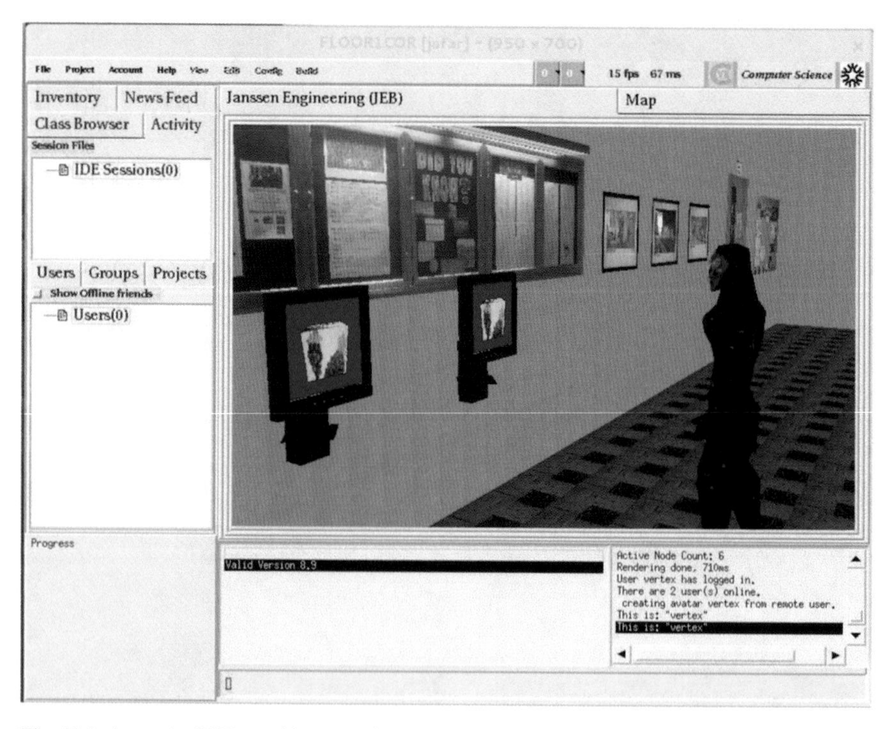

Fig. 11.1 A user in CVE watching two virtual computer screens sharing the same dynamic texture

board in the above example addressable as a visualization target over the network, it can be modified to read:

```
Rect { # whiteboard
   texture whiteboard.gif
   id 314
   coords [29.3,1,2.5, 29.3,2.5,2.5, 29.3,2.5,.4, 29.3,1,.4]
   }
```

The CVE clients maintain a table named gid that maps such global id's to texture records in the display list, with which the incoming message

```
\rpc EraseArea(w_314, 0, 0, 512, 512)
```

is interpreted by code that executes the equivalent of

```
EraseArea(gid[314], 0, 0, 512, 512)
```

11.2 Distributed Clones

Visualization tools often create multiple references with distinct drawing attributes for drawing on a particular drawable window (or in this case, texture). The X Window and OpenGL terminology is "graphics context", but in Unicon the more colorful term "clone" is used. Clones are a good example of entities requiring run-time assignment of global identifiers. In response to a user message of

```
\rpc Clone(w_314, "fg=red")
```

The server transmits

```
\gid 317
```

to the originator, assigning a new global id to the newly created resource. Instead of receiving a forwarded copy of the message, the other clients receive the aggregated message:

```
\rpc w_317 := Clone(w_314, "fg=red")
```

11.3 Control Flow and Time Warp

Normally, a program that is running is so much faster than the humans around it that we appear to be standing still. Similarly, visualizations of dynamic execution behavior will often have to be slowed down in order for humans to observe what they convey. In fact, much of the time humans need to reverse the relationship and effectively stop the program execution so that we can observe it from infinite points of view until we understand it. Animated visualizations intrinsically fall into an exciting in-between area, in which program speed is non-zero relative to the human observer. Different humans in different circumstances will need different animation rates, so other than to pick sensible defaults, the main goal is to leave the user in control of the speed of execution.

CVE renders 3D scene updates in real-time, and controls the central flow of execution. But monitors' control flow revolves around advancing the program state to the next event. Monitors resolve the duality of processing events from both program execution advances and user interactions by always giving user interactions priority, and stopping or slowing down execution as needed in order to leave users in control. Monitor coordinators have the same issue, with the additional understanding that running at full speed – that is, advancing the program state at a selected rate instead of single stepping – different visualizations' animations will run at different apparent speeds, proportional to the differences in frequency of their selected program execution events.

Since CVE is now a monitor coordinator, we can characterize its control flow and event processing behavior. CVE's control flow is controlled by an object called the dispatcher that embodies the event-driven paradigm. The target program is running on one of the CVE clients at a speed controlled by a slider that ranges from 0 (stopped) to 1 (as fast as possible). Control flow priority always goes to the user input. When running at full speed, the CVE client calls `EvGet()` whenever CVE's dispatcher does not have user or network input to process. The pseudocode of the dispatcher's method `message_loop(mainwindow)` looks like:

```
method message_loop(mainwindow)
   ...
   while mainwindow.is_open do {
      if there is user or network input then process it
      else {
         do_validate() | do_ticker() | do_nullstep() | do_monitorstep() | FPS < 35 | delay(idle_sleep)
         }
      }
end
```

The method `do_monitorstep()`, which advances a clock and may call `EvGet()`, implements the monitor coordinator functionality in a piecewise fashion. It is only called when the virtual environment does not have something else it has to do on behalf of the user. Figure 11.2 shows an example of two visualizations inside CVE.

Fig. 11.2 Two 3D visualizations runing at two different speeds inside CVE

11.4 Bug Hunting

This section is about turning bugs into monsters in the virtual environment, so that developers see them and are motivated to fix them. The point of looking at bugs in a book about software visualization is: users will often use visualization tools to help identify, locate, and fix bugs. The better the virtual environment is at making bugs visible and enabling the user to trace them to their causes, the more effective the tool will be. Topics in this section include

- Program bug reports as spawn points
- Bugs as clues about what to look for in a software visualization
- Different kinds of bugs, different ways of combating them
- Slaying bugs versus fixing them

Several aspects of this topic are research problems in their own right, and beyond the scope of this book. The focus of the section will be restricted to a particular illustrative example.

The virtual environment cannot render bugs unless they are reported. There are several sources of bug reports. The most reliable are explicit reports, posted to bug trackers such as Bugzilla, Traq, or Source Forge's Bug Tickets. For example, on 9/22/2014, there were 194 bugs listed at sourceforge.net/p/unicon/bugs/. Some fraction of them are fixed, but a good number are unresolved. In addition to those bugs, which are explicit and public, but easy to forget about if they are not happening to you, there are a number of channels that are less amenable. E-mails, telephone calls, postcards, or personal experience may reveal bugs not on the bug tracker. Even the software visualization tools described in, or developed as a result of, this book may reveal bugs that previously were unobservable.

Wherever your bugs came from, the primary goal of the virtual environment is to make them visible and motivate their eradication. It will help to make them annoying. Especially those bugs that the developer is not seeing when they run their own tests need a visible representative, an avatar, to represent and draw attention to themselves.

Some bugs are found and resolved quickly, but those that are not need to call attention to themselves periodically. Like the ghost of Hamet's father, or many named monster NPC's in world of warcraft, the bug should be thought of as a spawn point that creates a visible avatar every so often. High priority bugs need more attention and should spawn more often.

Inside the virtual environment, bugs may be dismissed temporarily. Low priority bugs might be dismissed simply by clicking on them, or by executing a videogame-style combat sequence. Higher priority bugs need to be more persistent, so they may require that a quest be fulfilled before they go away. Like an unwanted mariachi band, you might have to tip them to go visit some other table for a few minutes.

Appendices

Appendix A: Event Codes

The event codes provided in the Unicon runtime system are presented below. The nature and extent of this instrumentation is discussed in Chap. 3.

A.1 Classes of Events

AllMask	The universal set, all events
AllocMask	Memory allocation
AssignMask	Assignment
ConvMask	Type conversion
EmptyMask	The empty set
EvalMask	Function, procedure, and operator activity
FncMask	(Built-in) Function activity
ListMask	List operation
OperMask	Operator activity
ProcMask	Procedure activity
RecordMask	Record operation
ScanMask	String scanning
SetMask	Set operation
StructMask	Structure operation
TableMask	Table operation
TypeMask	Types as in AllocMask, plus integer, null, and proc

C. Jeffery, J. Al-Gharaibeh, *Writing Virtual Environments for Software Visualization*, DOI 10.1007/978-1-4614-1755-2

A.2 Individual Events

Code	Value	Description
E_Aconv	Input value	Conversion attempt
E_Alien	# bytes	Alien allocation
E_Aloc	Line/column #	Location change (artificial)
E_Assign	Variable name	Assignment
E_BlkDeAlc	# bytes	Block deallocation
E_Coact	Co-expression	Co-expression activation
E_Coexpr	# bytes	Co-expression allocation
E_Cofail	Co-expression	Co-expression failure
E_Collect	Region number	Garbage collection
E_Coret	Co-expression	Co-expression return
E_Cset	# bytes	Cset allocation
E_Cstack	# bytes	C stack depth
E_Deref	*Varies*	Dereference variable to obtain its value
E_Disable	*Varies*	Disable monitoring (artificial)
E_Enable	*Varies*	Enable monitoring (artificial)
E_EndCollect	Null	End of garbage collection
E_Error	Error number	Run-time error
E_Exit	Exit code	Program exit
E_External	# bytes	External allocation
E_Fcall	Function	Function call
E_Fconv	Input value	Conversion failure
E_Ffail	−1	Function failure
E_File	# bytes	File allocation
E_Free	# bytes	Free region
E_Frem	0	Function suspension removal
E_Fresum	0	Function resumption
E_Fret	Result	Function return
E_Fsusp	Result	Function suspension
E_HashNum	Integer	Hash number
E_HashSlots	Integer	Number of slots in the hash table
E_HashChain	Integer	Length of hash chain traversed within the slot
E_Intcall	Interpreter signal	Interpreter call
E_Integer	Result	Integer value pseudo event
E_Intret	Interpreter signal	Interpreter return
E_Lbang	List	List generation
E_Lcreate	List	List creation
E_Lelem	# bytes	List element allocation
E_Lget	List	List get, same as E_Lpop

Code	Value	Description
E_Line	Line #	Line change
E_List	# bytes	List allocation
E_Literal	*Varies*	Push a literal (constant) value onto the stack
E_Loc	Line/column #	Location change
E_Lpop	List	List pop, same as E_Lget
E_Lpull	List	List pull
E_Lpush	List	List push
E_Lput	List	List put
E_Lrand	List	List random reference
E_Lref	List	List reference
E_Lrgint	# bytes	Large integer allocation
E_Lsub	Subscript	List subscript
E_MXevent	Window event	Monitor input event
E_Nconv	Input value	Conversion not needed
E_Null	Null	Null value pseudo event
E_Ocall	Operation	Operator call
E_Ofail	−1	Operator failure
E_Opcode	Operation code	Virtual-machine instruction
E_Operand	Operand	Virtual machine operand
E_Orem	0	Operator suspension removal
E_Oresum	0	Operator resumption
E_Oret	Result	Operator return
E_Osusp	Result	Operator suspension
E_Pcall	Procedure	Procedure call
E_Pfail	Procedure	Procedure failure
E_Prem	Procedure	Suspended procedure removal
E_Presum	Procedure	Procedure resumption
E_Pret	Result	Procedure return
E_Proc	Procedure	Procedure value pseudo event
E_Psusp	Result	Procedure suspension
E_Rbang	Record	Record generation
E_Rcreate	Record	Record creation
E_Real	# bytes	Real number allocation
E_Record	# bytes	Record allocation
E_Refresh	# bytes	Refresh allocation
E_Rrand	Record	Record random reference
E_Rref	Record	Record reference
E_Rsub	Subscript	Record subscript
E_Sbang	Set	Set generation
E_Sconv	Output value	Conversion success

Code	Value	Description
E_Screate	Set	Set creation
E_Sdelete	Set	Set deletion
E_Selem	# bytes	Set element allocation
E_Set	# bytes	Set allocation
E_Sfail	Old subject	Scanning failure
E_Signal	Signal name (string)	
E_Sinsert	Set	Set insertion
E_Slots	# bytes	Hash header allocation
E_Smember	Set	Set membership
E_Snew	New subject	Scanning environment creation
E_Spos	Position	Scanning position
E_Srand	Set	Set random reference
E_Srem	Old subject	Scanning environment removal
E_Sresum	Restored subject	Scanning resumption
E_Ssasgn	Length of result	Sub-string assignment
E_Ssusp	Current subject	Scanning suspension
E_Stack	Stack depth	Stack depth
E_StrDeAlc	# bytes	String deallocation
E_String	# bytes	String allocation
E_Sval	set element	Set value
E_Syntax	Control structure name	Control structure entry/exit
E_Table	# bytes	Table allocation
E_Tbang	Table	Table generation
E_Tconv	Example target	Conversion target
E_Tcreate	Table	Table creation
E_Tdelete	Table	Table deletion
E_Telem	# bytes	Table element allocation
E_TenureBlock	Size	Tenure a block region
E_TenureString	Size	Tenure a string region
E_Tick	# ticks	Clock tick
E_Tinsert	Table	Table insertion
E_Tkey	Table	Table key generation
E_Tmember	Table	Table membership
E_Trand	Table	Table random reference
E_Tref	Table	Table reference
E_Tsub	Subscript	Table subscript
E_Tval	Table element	Table value
E_Tvsubs	# bytes	Substring trapped variable allocation
E_Tvtbl	# bytes	Table element trapped variable allocation
E_Value	Value assigned	Value assigned

Appendix B: Software and Supporting Documentation

Unicon lives at unicon.org. Unicon is described in a book by Jeffery et al. [1]. It is a superset of the Icon language. The graphics facilities used are described extensively by Griswold et al. [2], and by Martinez et al. [3].

References

1. C. Jeffery, S. Mohamed, R. Parlett and R. Pereda, Programming with Unicon, 2003.
2. R. E. Griswold, C. L. Jeffery and G. M. Townsend, Graphics Programming in Icon, San Jose, California: Peer-to-Peer Communications, 1998.
3. C. Jeffery, N. Martinez and J. Al-Gharaibeh, "Unicon 3D Graphics: User's Guide and Reference Manual (Unicon Technical Report #9b)," 2010.

Printed by Printforce, the Netherlands